Star Trek on the Brain

STAR TREK ON THE BRAIN:

ALIEN MINDS, HUMAN MINDS

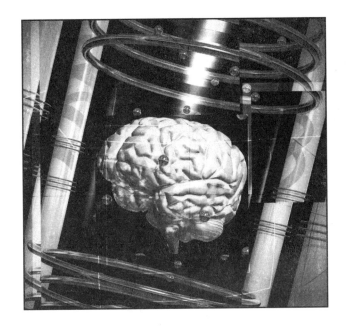

Robert Sekuler
Randolph Blake

W. H. Freeman and Company
New York

Cover Design: Roy Weimann
Text Design: Victoria Tomaselli
Composition: Bonnie Stuart
Illustrations: Laura Duprey

Star Trek is a registered trademark of Paramount Pictures.

Library of Congress Cataloging-in-Publication Data

Sekuler, Robert.
Star Trek on the brain: alien minds, human minds / Robert Sekuler,
Randolph Blake.
p. cm.
Includes index.
ISBN 0-7167-3279-3 (hardcover)
ISBN 0-7167-3692-6 (paperback)
1. Neurosciences. 2. Star Trek television programs. I. Blake, Randolph.
II. Title.
RC341.S39 1998
612.8'2—dc21   98-9433
                                                                    CIP

Printed in the United States of America

First paperback printing, 1999

For Susan Sekuler
and Elaine Blake

Contents

Preface

Authors' Log
Earthdate 1/18/98

For over thirty years, the *Star Trek* television series and motion pictures have dazzled viewers with a rich and complex universe of aliens and alien cultures. But these aliens and their worlds are much more than continuing sources of entertainment: They are metaphors for ourselves and *our* world. Over the course of *Star Trek's* history, these metaphors have at times taken the form of creatures with hyperlogical minds, creatures who can read minds, creatures made of silicon, creatures who believe sex should be illegal, and creatures whose genetically engineered drug addiction makes them natural born killers. These and other equally bizarre *Star Trek* aliens provide excellent case studies from which we can learn important lessons. With open minds and a little imagination, we realize that these aliens are really not all that different from ourselves, one of the major themes of this book. And it's not just *Star Trek's* aliens who provide useful case studies—*Star Trek* also paints an entertaining and provocative portrait of our *own* species three centuries from now, giving us an imaginative glimpse of how our future relatives may think, feel, and act.

Some of our friends haven't watched much *Star Trek*, and perhaps you haven't either. So we'll give you the same assurance we gave them: You don't need to be a seasoned Trekker to enjoy this book and appreciate its lessons! Don't worry if you're not familiar with the characters from or details of the television series episodes and motion pictures. As long as you understand the basic premise of *Star Trek*—humans in the future exploring galaxies filled with weird and wonderful aliens—the rest will fall into place.

The approach we took in this book isn't novel. In fact, it borrows from what we do in the courses we teach in neuroscience and psychology. There we often illustrate for students how a normal human brain operates by showing what happens when an accident (for example, a stroke) or a genetic abnormality (for example, a missing enzyme) produces something other than a normal brain. And our experiences with such illustrations show that a knowledge of brain dysfunction helps students better understand how a normal brain works to control thought, emotion, and behavior.

Scientists have long taken a similar approach. For decades, biologists and psychologists have studied the physical structures and behavioral adaptations of other species, often to learn more about our own. For example, consider *Anableps anableps*. This strange South American fish has four eyes—two for underwater use and two for above-water use. A study of its vision may well reveal something about human vision. Or consider chimpanzees. These close primate relatives of ours can learn vocabularies consisting of several hundred "words" and then use those vocabularies to communicate with other chimpanzees as well as with humans. By studying chimpanzee conversations, we can likely add to our knowledge of human language.

There's no doubt that studies of other species enrich our understanding of biology and psychology. However, we believe that psychologists are setting their sights too low by restricting their inquiries to terrestrial creatures. After all, why should we limit ourselves to examining only *earthly* minds and behavior? We propose enlarging psychology's domain of study to include planets other than our own, thereby creating a science that might be called *exopsychology*—literally, the psychology of life outside Earth. Exopsychology's mission? To examine the minds and behavior of creatures from all over the universe.

Star Trek advances this mission by providing an abundant source of case histories involving alien brains and minds—such as those of Klingons, Vulcans, Ferengi, Ocampa, and the Horta (one of our favorite *Star Trek* life-forms). We of course realize that the television series and motion pictures that bear its name are fiction. But we believe fiction can be as powerful as reality, and these *Star Trek* case histories furnish deep insights into the brains and minds of *Homo sapiens*, as you will see.

A word about our sources. We went to great lengths to get the science in this book right. Toward this end, chapters are filled with numerical references to scientific writings that have guided our thinking. Descriptions of these references are included in the Notes section at the end of the book. Furthermore, preliminary versions of most parts of the book were read by experts whose comments and corrections shaped our revisions.

We worked equally hard to get the *Star Trek* part of it right, too. Many devotees of *Star Trek* know its characters, television series episodes, and motion pictures inside out, and to serious Trekkers, even the tiniest of mistakes would stand out like a bald Talaxian. And so, to be certain that our treatment of *Star Trek* would be as accurate as possible, we watched its television series episodes over and over again, checking that we'd transcribed dialogue correctly and interpreted plot nuances clearly. For this portion of our research, we made extensive use of our treasured collections of *Star Trek* videocassettes, which include every episode of all four series, plus all eight movies.

From our research, teaching experience, and personal experience, we are well aware that human memory can be treacherously fallible. Therefore, to check facts and aid our own failing memories while writing *Star Trek on the Brain*, we consulted various books on *Star Trek*. These included Michael and Denise Okuda's *The Star Trek Encyclopedia* and Phil Farrand's excellent *The Nitpicker's Guide* series. Also of great comfort to us were Larry Nemecek's *Star Trek: The Next Generation Companion* and Hal Schuster and Wendy Rathbone's *Trek: The Unauthorized A–Z*. In addition to these books, the World Wide Web proved to be a veritable latinum mine of information on *Star Trek*. Particularly excellent were the plot summaries of *Star Trek* television series episodes contained in David R. Landis's *The Star Trek Episode Guide* HyperCard™ stacks; these stacks are available as freeware at **http://www.universe.digex.net/~oakmtsw/**.

The two of us have worked together on and off for over two decades. During that time, we've collaborated on three editions of another book, on magazine articles, on scientific research projects, and in the classroom. All those collaborations were fun, but for sheer pleasure, this book dwarfs everything that came before it. Writing it has been *the real fun*. Part of our pleasure in working on this project came from the enthusiasm and support we received from many, many people around the world who are bound by their love of *Star Trek*. They helped us understand the true universality of *Star Trek's* appeal. And so, as this book nearly became a group project, we have more than the usual number of people to thank.

Erica Beth Sekuler, Allison Britt Sekuler, and Brian Lebovitz were instrumental in helping us formulate the book and in identifying *Star Trek* television series episodes that would illuminate many of the topics we wanted to cover in it. They were there at the beginning and motivated us to believe that the book was actually doable.

Marilyn Baker and Steve Marbit lent us videocassettes, pointed us to key television series episodes, and spent hours talking *Star Trek* with us. Micky Josephs, John Hummel, Cynthia Mark-Hummel, Maja Mataric, Richard Roberts, Warren Craft, and Stacia Sekuler Miehe provided other *Trek*-related inspirations and ideas.

Carolyn Cave, Jo-Anne Bachorowski, William Caul, Leslie Smith, and Geoffry Hurst read and commented on draft versions of chapters.

Three people generously supplied some of the images contained in the book: William Whetsell (photographs of brain tissue), Robert Kessler (a PET image), and William A. Kimmins (photographs of facial expressions associated with emotions, for which Paul Soper struck very accurate poses).

We thank Robert Wurtz, Stanley Finger, Elizabeth Loftus, Haruyuki Kojima, Adar Pelah, Merlin Butler, Wendy Stone, Jon Kaas, Irwin Levitan, and Aryeh Routtenberg, all of whom graciously gave us technical help with scientific questions.

We are grateful to Elizabeth Knoll, our supporter and first editor, to Susan Finnemore Brennan, our second editor, to Norma Roche, who did a spectacular job of copyediting our manuscript, and to Christopher Miragliotta, our project editor.

Our loving and supportive wives, Susan Sekuler and Elaine Blake, suffered patiently our preoccupation—OK, *obsession*—throughout the entire project. And when that faraway look came into our eyes, Susan and Elaine understood that, while our bodies were on Earth, our minds were in a distant part of the galaxy. They also had faith in us—for instance, they believed that, during the times we sat on the couch watching *Star Trek* and drinking Samuel Adams (we couldn't get any Romulan ale), we were actually engaged in serious research. Furthermore, they read and commented on draft versions of chapters. For all these things, Susan and Elaine deserve the highest commendation the United Federation of Planets can award to civilians, but for the moment, they'll have to settle for a simple and heartfelt thanks.

One final note before we launch our "star trek" through the brain. The mission on which we've embarked is a *continuing* one—it did not end with our completion of this book. With luck, we'll have an opportunity to continue this trek through the brain and tap some of the material and ideas that didn't make it into these pages. Toward this end, we welcome your comments and suggestions. We can be reached by e-mail at **sekuler@volen.brandeis.edu** and **randolph.blake@vanderbilt.edu**. Finally, we have set up a site on the World Wide Web that contains material related to this book. There you'll find quizzes, additional insights into human minds and alien minds, and—most important of all—reader comments and suggestions as well as our replies to them. The Web site's Internet address is: **http://www2.shore.net/~sek/STonthebrain.html**.

Star Trek on the Brain

In the twenty-fourth century, rank in Starfleet is signified by pips, small metal pins that are worn on the uniform collar. Jean-Luc Picard, captain of the United Star Ship *Enterprise*-D, wears four gold pips. So do Kathryn Janeway, captain of the U.S.S. *Voyager*, and Benjamin Sisko of space station Deep Space Nine.[1]

In twentieth-century Colorado, Dr. Temple Grandin sometimes wears just two pips, and their design is definitely not regulation Starfleet. Her pips are silver pins in the shape of cattle. They testify to her unique gifts as a creator of equipment for feeding and handling livestock. Grandin, a professor at Colorado State University, has a unique empathy for these animals that enables her to design systems that set industry standards for humane treatment.[2]

Like many people, Grandin has been a fan of *Star Trek* from its inception three decades ago. Her debt to *Star Trek* is complicated and fascinating. For her, the show has been not just entertaining, but also therapeutic. At one especially stressful time in her life, an hour each afternoon with *Star Trek* calmed her down and reduced her anxiety attacks.[3]

Temple Grandin is an entrepreneur, inventor, scientist, and author of several well-received books. She also has autism, one of the human brain's most misunderstood and bewildering disorders. It affects about one child in every thousand and, as far as we know, lasts for the individual's entire life. Many people know autism only through Dustin Hoffman's character in the movie *Rain Man*, but that character shows just one variety of autism. The disorder can take many guises, spanning an enormous range of severity. Always, though, autism constricts the affected person's interests and activities, and impairs social and communication abilities.

Temple Grandin is a high-functioning autistic. Her autism carries outward signs so subtle that the condition can go undiagnosed or misdiagnosed. She has a wonderful vocabulary, but little or no intuitive sense of how her words affect others. She cannot feel the feelings that motivate most human beings. *Star Trek* helped Temple Grandin appreciate how she was different from other people.

> *I vividly remember one old episode because it portrayed a conflict between logic and emotion in a manner I could understand. A monster was attempting to smash the shuttle craft with rocks. A crew member had been killed. Logical Mr. Spock wanted to take off and escape before the monster wrecked the craft. The other crew members refused to leave until they had retrieved the body of the dead crew member. To Spock, it made no sense to rescue a dead body when the shuttle was being battered to pieces. But the feeling of attachment drove the others to retrieve the body so their fellow crew member could have a proper funeral. It may sound simplistic, but this episode helped me finally understand how I was different. I agreed with Spock, but I learned that emotions will overpower logical thinking, even if these decisions prove hazardous.[4]*

Temple Grandin is not the only person who can learn important things from *Star Trek*. It holds lessons aplenty, if you know where to look for them. By thrusting us vicariously into novel, atypical situations, *Star Trek* challenges our habitual ways of thinking. By bringing viewers into close contact with all sorts of nonhuman life-forms and their cultural customs, *Star Trek* forces us to see ourselves in a new, revealing light. That light is often illuminating, if sometimes a bit too bright for comfort.

> *Who cares about anomalies? People want stories about things they can relate to. Life and Death. Good and Evil . . .*
>
> Jake Sisko, writer and son of Starfleet officer Benjamin Sisko
> (DS9-5, "Nor the Battle to the Strong")*

To say that *Star Trek* is about space exploration is like saying that sex is about making babies. Both statements contain a kernel of truth, but each misses the point. What began thirty years ago as a television series with low-budget makeshift props, improvised not-so-special effects, and often predictable dialogue has become a part of the world's cultural landscape. It has deepened our awareness of our place within a larger universe, and has given us imaginative glimpses of where space travel and the future might take us. But *Star Trek* has also deepened our awareness of who we are, right here on Earth.

Star Trek's creator, Gene Roddenberry, was its guiding spirit until his death in 1991. Together with talented writers, directors, and actors, Roddenberry helped give life to archetypal characters whose challenges in the twenty-third and twenty-fourth centuries are remarkably similar to those we face today. *Star Trek's* various incarnations have captured the hearts, minds, and imaginations of millions of people all over the planet, overcoming barriers of language, race, time zones, and even interruption by commercials.

If special effects were all that *Star Trek* had had going for it, the idea would have been truly short-lived, a three-years-and-out blip in television history.[5] But there has always been much more to *Star Trek* than the visuals. Devotees and occasional viewers alike have been captivated by *Star Trek's* portrayals of human beings, humanoids, and utterly alien beings who face challenges that are at once exotic and familiar. Even when *Star Trek* seems to focus on some weird alien life-form, the story is really about *Homo sapiens*. For example, a huge and beautiful crystalline entity is the focus of an entire episode (TNG-5, "Silicon Avatar"). We learn a lot about the entity's history and about the havoc it wreaks on innocent life-forms who cross its path. But the episode's lessons are really about the immense power of a mother's love for her child, and about the senseless revenge she takes on her child's killer. In another episode, a powerful creature called the Dal'Rok terrorizes Bajoran villagers every year during their harvest season (DS9-1, "The Storyteller"). Only by combining their voices are the villagers able to shout loudly enough to scare away the Dal'Rok. On its surface, the episode is about a strange-looking cloud that scares some unsophisticated farmers. But at a deeper level, the story deals with the power of myth to shape a people's collective consciousness and unite a group in common purpose.

Star Trek's human dimension holds up a mirror in which we can see ourselves. Sometimes that mirror reflects images not entirely different from those we have observed on Earth: gangster-ridden Chicago in the 1920s (TOS-1, "City on the Edge of Forever"), Hitler's Germany in the 1930s (TOS-2, "Patterns of Force"), or Roswell, New Mexico, in 1947 (DS9-4, "Little Green Men"). Other times, *Star Trek's* mirror does not just reflect, but magnifies human characteristics in a way that awakens us to things about ourselves we might ordinarily overlook. *Star Trek's* scenarios are not really about avaricious and scheming Ferengi, hyperlogical Vulcans, or boastful Klingons. Those characters are us, and their qualities are ours. Ferengi avarice is merely an exaggerated, more persistent form of the avarice that nearly all humans experience to one degree or another. Ditto for Vulcan logic and Klingon boastfulness.

Star Trek often uses humor to make important points about human characteristics. For example, it can poke fun at something as universal as our tendency to see what we want to see. Remember lying on your back, looking up at the clouds overhead, and making up stories about what objects and creatures those clouds resembled? Guinan and Data, neither of whom is human, are on board the *Enterprise*-D playing a version of this game (TNG-5, "Imaginary Friend"). As they look out the window, they see a remarkable, huge, reddish swirling thing. Neither one knows what it is, but each behaves true to form. Guinan, with great imagination, first envisions a beautiful coral fish, then a Mintonian sailing ship. Data, the coolly analytical android, is unable to see these forms. He concludes that humanoids tend to see meaningful patterns even when they're looking at completely random nonsense.

Guinan is not *Homo sapiens*, but she seems to perceive as we do. In the random swirls she sees exactly the kind of structure you'd expect of a brain that uses some of the same neural machinery for both perceiving and imagining. And that's precisely what the human brain does. Our compulsion to squeeze sense out of nonsense is a necessary and valuable side effect of the way in which our brains process sensory information.

The blurry swirls in the pictures on the facing page are not as beautiful as the swirls that Data and Guinan so admired, but they are interesting nonetheless. If you look at the swirls in the left-hand picture, you won't see a Semalian coral fish or a Mintonian sailing ship, but there is a good chance that after a while you will see a person's face (hint: look in the upper left-hand quarter of the picture). In the right-hand picture, you may see another face (hint: look in the upper right-hand quarter of the picture). If you don't see the faces right away, keep looking. For some people they take a while to materialize, but once they do register, it'll be hard not to see them. The amazing part? Your brain creates them out of meaningless blobs. (At the end of the chapter you'll find a short explanation of how these pictures were made.)

 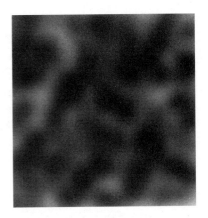

Imagining the Imaginable

Like works of science fiction in any medium, *Star Trek* speculates on as yet unrealized, but plausible, scientific developments. Of course, not all of its speculations are equally plausible. Take the radical neurosurgery that is performed on Spock in the television series episode "Spock's Brain" (TOS-3). It's hard to imagine how Spock's entire brain could be plucked out of his head without damaging his skull or even mussing his hair. It's even harder to understand how, later, the brainless Spock is able to instruct the surgeon as his brain is being reinstalled. Late-twentieth-century neuroscientists are working hard on ways of encouraging nerve regrowth.[6] Their early successes show that it is not total fantasy to dream about complete recovery from spinal cord damage and maybe, someday, even severe trauma to the brain. It's doubtful, however, that the procedure performed on Spock will ever find its way into neurosurgery's arsenal. But *Star Trek* doesn't have to be perfectly plausible—it is fiction, after all, not reportage. Despite its occasional forays into fantasy, *Star Trek* offers sound, valuable lessons about real brains and the real minds those brains create.

Various incarnations of the U.S.S. *Enterprise* have explored our galaxy, the Milky Way, and occasionally wandered beyond, where no one has gone before. These explorations depended upon the invention of fictional devices by *Star Trek's* writers—warp drives, transporters, replicators, and the like. For students of physics, *Star Trek* offers up a melange of the plausible, as well as the impossible: things that could not happen unless the laws of physics were repealed.[7] Those aspects of *Star Trek* represent one stream of science fiction. To many people, though, the most interesting dimension of *Star Trek* lies not in its focus on science, but in what Isaac Asimov called "social science fiction," its concern with how scientific advances affect human beings.[8]

Our own fascination with *Star Trek* is not with its technology and special effects. Instead, we're intrigued by its thought experiments, its "what ifs" about people and society. As good fiction always does, *Star Trek* lifts us out of the world we inhabit every day and thrusts us into new situations that challenge our assumptions and habits:

• Suppose some humanoid society managed to eradicate all biochemical traces of differences between males and females. In such a society, what would happen to an unlucky misfit born with a birth defect that made it male or female?

• Suppose technology made it possible for a society to link together the minds of all its people. If the result were a harmonious group consciousness and a shared, cohesive memory, how would these linked beings view us pathetic nonlinked creatures who must remember, think, decide, and act all alone, one individual at a time?

• Suppose technology let you simulate, ahead of time, any social interaction that you might need to take part in. Using a holodeck, you could practice for all the difficult moments of your life: romantic encounters, asking the boss for a raise, disciplining a teenage child, talking your way out of a traffic ticket. If you could practice any of these interactions over and over, how would your simulated experiences affect the real thing?

It's not surprising that so many people connect so strongly with *Star Trek's* stories, even though they take place in distant parts of the universe, on starships travelling at warp speeds. No matter how clever the writers and directors are, they are still bound by natural constraints on the human imagination. In crafting stories about *Star Trek's* crew and the alien beings they encounter, *Star Trek's* writers don't create their fictional worlds out of whole cloth: they draw on things that they—and we—know.

And this is the root of all creativity: a novel rearrangement of ordinary, familiar things. When it was first shown in public in 1907, Pablo Picasso's painting *Les Demoiselles d'Avignon* was hailed for its revolutionary approach and reviled for its break with tradition. The painting's five distorted female figures with their strange masklike faces struck viewers as utterly unlike anything they'd ever seen before. Though many critics now consider it the most important painting of the twentieth century, it seemed alien and offensive to many people. But every element of the painting has a clear antecedent in things Picasso had actually seen. Preliminary sketches for the painting, x-ray studies of its various stages, and the study of various influences on Picasso all show that the painting was constructed out of elements that were known or even familiar to the artist.[9] In Picasso's masterpiece, those ordinary elements have been transformed into a whole that is startling in its novelty.

You may not be Picasso, but you can see something of this creative process in yourself. Take a few minutes to form a clear mental image of an extraterrestrial creature that inhabits some strange, distant planet. Assume that the alien's planet of origin is completely different from Earth, and that it has never before been seen by any humanoid. Now, before you read on, make a sketch of your imagined creature.

When people try this exercise, they tend to produce imaginary creatures that bear an uncanny resemblance to creatures found right here on planet Earth.[10] Their creations tend to have symmetrical left and right sides, limblike appendages, a head or two atop their bodies, and obvious external sensory organs. These "exotics" are assembled from familiar components, and their overall appearance, though imaginary, betrays their terrestrial roots. Your own newly minted alien might resemble your baby sister, or perhaps that horrid kid who tormented you in kindergarten. You are a human being, and when human beings try to imagine alien worlds and their inhabitants, the limits of our imagination come to the fore.[11]

The things we can imagine are deeply rooted in what we know and what we've experienced. That's one reason why *Star Trek's* episodes about love, hate, greed, prejudice, bravery, and other familiar human qualities resonate so deeply. The episodes that are most difficult to understand are the ones that push the envelope of viewers' imagination. In "Frame of Mind" (TNG-6), Commander William Riker shifts back and forth between several very different realities, making it difficult for us to keep track of what's real and what isn't. The same challenge faces us when we try to follow episodes in which characters travel backward or forward in time, sometimes altering the flow of history.[12] In these "alternative timeline" episodes we may even witness events whose details are determined by other events that have not yet occurred. When time runs outside its normal one-dimensional channel, our imaginations are stretched to their limits. Experience hasn't prepared us for time travel.

Hearts and Minds

If Aristotle had watched *Star Trek*, he undoubtedly would have been puzzled by a lot of what he saw: food replicators, transporters, warp drives, communication badges, holodecks.[13] But probably nothing would have puzzled him more than *Star Trek*'s obsession with the human brain. Aristotle believed that all rational functions were carried out in the human heart. He was right in one respect: mental thought is a biological function. But he identified the mind with the wrong body part. To Aristotle, the brain was nothing more than a radiator designed to keep the heart cool. Human brains were big, he reasoned, because the heat generated by the large human heart demanded a large radiator. In his view, the brain had nothing to do with intellectual function.

Aristotle was not the first person to treat the human brain with irreverence. In ancient Egypt, when priests prepared a body for mummification, they routinely scooped out the brain and threw it onto the trash heap. If the brain wasn't terribly important while the deceased person was alive, it certainly couldn't be of much use in the afterlife.[14] The heart, which the Egyptians, too, thought was the organ of mental function, was preserved with great care so that its temporarily dead owner would be able to use it on Judgment Day and beyond.

Nowadays, of course, we know better. The brain is where the action is, although you wouldn't know it from appearances alone. In truth, your brain is not much to look at. Its surface is light gray, deeply corrugated with valleys and ridges and crisscrossed with blood vessels. Who could believe that this wrinkled lump of tissue is at the center of our physical and mental lives? In fact, the brain's power is the stuff of which everything human is made. And among the brain's creations, none is more remarkable—or elusive—than the mind.

The mind is not something that you can hold in your hand—or put in a jar. "Mind" is a collection of amazingly diverse functions, including thought, emotion, and creativity. Planning and coordination and telling a joke are also the mind's doing. Running a race or playing the piano are just as much *mental* functions as are solving a complex problem in algebra or writing a poem. Every mental function, whatever its outward form, reflects the operation of the nervous system, particularly the brain. In fact, "minds are simply what brains do."[15]

Because this book is about brains and the minds they create, we will start with a brief overview of the human brain.

A Warp-Speed Tour of the Brain

The human brain is a lightweight, tipping the scales at a paltry 3 pounds, give or take a few ounces. For your average 160-pound person, that's only about 2 percent of total body weight. But size is not everything. According to one noted neuroscientist, "The human brain is by far the most complex structure in the known universe."[16] This lump of soft, squishy tissue is a device of staggering intricacy and power. It contains about 100 billion nerve cells, called *neurons,* many of which look something like the three shown in the illustration below. The many branches projecting from the cell body of each neuron allow it to communicate with thousands of its

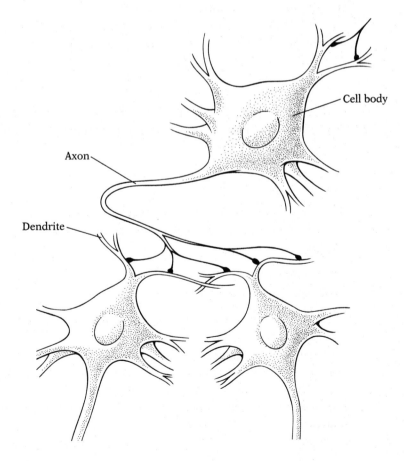

Cell body

Axon

Dendrite

neighbors. One branch, called an *axon*, transmits signals from the cell body. These signals are then received by a set of branches called *dendrites*. The complexity of this communication network empowers the brain's owner with unimaginable capabilities.

In the brain's interior, neurons form many overlapping confederations, which can compete or collaborate with one another. Each confederation of neurons can be defined by the jobs it performs (function) and by the specific region within the two hemispheres of the brain where it is located (anatomy). Some groups of neurons control the body's muscles, allowing us to walk, type, talk, move our heads and eyes. We call these *motor neurons* in recognition of their role in guiding our motor activities. Other groups of neurons decipher sensory messages received by the eyes, ears, and other sensory receptors scattered over the body's surface. Without these *sensory neurons* we'd be cut off perceptually from the external world. Still other groups of neurons monitor and regulate the body's internal state, including its temperature, blood oxygen concentration, heart rate, and hormone levels. These neurons signal when it's time to eat, sleep, slip on a coat, or search for a sexual partner.

Finally, a huge proportion of the human brain contains specialized groups of neurons devoted to higher cognitive functions such as language, reasoning, and decision making. These specialized brain regions are important in emotion, memory, aggression, perception, and reasoning. They give us insight into many important questions about ourselves:

• What makes it possible to remember—or forget—a lifetime's worth of experiences?

• What are the origins of our aggressive tendencies?

• What's sex for, and how come it's so much fun?

• What goes wrong when someone gets depressed or becomes demented?

• Does everyone experience the same emotions?

Every one of the brain's specialized regions has two things in common. First, everywhere in the brain, neurons create tiny electrical signals, called *action potentials*. When a neuron produces a burst of these electrical pulses, we say that the neuron is active. Because there are so many neurons, each individual neuron's action potentials make an infinitesimal contribution to the overall neural give-and-take that makes up brain activity. What counts most is the ever-changing pattern of neural activity throughout the brain. In this respect, different specialized groups of neurons behave like regiments within this large army of brain areas. Each has a particular responsibility, and all work cooperatively toward a common goal: enabling the brain's owner to survive and flourish in a complex physical and social environment.

The second important property of neurons comes out of their ability to communicate with one another. To do their jobs, neurons must interact. This interaction is made possible by *neurotransmitters*, chemicals that pass messages from neuron to neuron. By the latest count, the human brain makes use of several dozen different neurotransmitters. Frequent mention in the news media has made a few of these neurotransmitters household names, particularly serotonin and dopamine. You may have read about drugs, such as Prozac, that work by boosting the effects of serotonin. Or you may have read that dopamine is being implicated in a variety of addictions.

All neurons secrete tiny amounts of neurotransmitter, and these chemical substances quickly float across the tiny liquid-filled gap, or *synapse*, that separates each neuron from its neighbors. These gaps are typically about 20 nanometers (20 billionths of a meter) wide, so small that more than 100,000 synapses could be stacked side by side quite comfortably within the letter *o*. After they have diffused across a synapse, molecules of neurotransmitter are picked up by specialized proteins (*receptors*) embedded in the membrane of the receiving neuron. This pickup of molecules triggers a cascade of chemical events that ultimately change the receiving neuron's electrical state and determine whether or not it will produce action potentials. Some neurotransmitters, called *excitatory neurotransmitters*, increase the likelihood that the receiving neuron will be activated. Other neurotransmitters, called *inhibitory neurotransmitters*, decrease the likelihood that the receiving neuron will be activated. From start to finish, these chemical events take a few thousandths of a second, far less time than the blink of an eye. This speed allows them to be repeated over and over, hundreds of times in a second. Any given neuron, then, can contribute many action potentials to the symphony of activity that is being played out by the entire brain.

Neurons don't just pass messages back and forth like so much biochemical e-mail. If they did, the brain wouldn't be able to do the complicated and interesting things it does. It would simply react reflexively to sensory events in the environment. But the brain that invented Chinese cooking and MTV is obviously much more unpredictable than this. That unpredictability comes out of interactions among hundreds of thousands of neurons. Working together, groups of neurons can carry out calculations, make decisions, store information, and create conscious awareness. How are these things accomplished by neurons that can do nothing more than generate electrical signals?

Any one neuron's activity level can be seen as a "decision" reached by weighing the excitatory and inhibitory "evidence" collated from thousands of other neurons in synaptic contact with it. Each neuron receives both kinds of evidence and weighs them like a poll-wise politician who is running for reelection: the neuron tallies up all the messages it receives, and crafts its own message accordingly. This process is going on throughout the entire brain, among all its constituent neurons, all the time.

When we take into account both the total number of neurons in the human brain *and* the number of synaptic connections each one makes with others, we begin to get a glimpse of the brain's true complexity. The number of connections among the brain's neurons is staggering: about 10^{14}! (At the risk of redundancy, that number is 100,000,000,000,000; translating all those zeroes into words, it's one hundred trillion.) No computer on Earth comes anywhere close to this level of complexity. What makes the brain's power even more mind-boggling is that the strength of connections among neurons can change over time with experience—the connections are *plastic*. (No, not like your coffee cup.) These changes in the strength of synaptic connections constitute what we call learning. Neural circuits are being modified throughout your life, enabling you to develop and draw on new skills and information. Your brain's plasticity lets you adjust—complain, but adjust—when the phone company changes your area code for the umpteenth time. Chief engineers aboard Starfleet vessels use their brains' plasticity to deal with updates in the ship's navigational and propulsion systems and to learn the professional strengths and personality quirks of new personnel.

Our Continuing Mission

The human brain has been a long time in the making—about 3 billion years. With that brain, we've managed to create cultures, invent hundreds of languages, explore space (though not as far out as the *Enterprise* goes), build weapons of mass destruction, write symphonies, break the genetic code, and create *Star Trek*. We don't know what new frontiers the human mind and brain will conquer during the coming centuries, but *Star Trek* holds out some provocative and challenging possibilities (if only because early-twenty-first-century engineers and astrophysicists will have grown up with ideas from *Star Trek*).[17]

The chapters to come are a trek through human minds and alien minds, through human brains and alien brains. Each chapter has a pair of missions: first, to explore some lessons about the mind that *Star Trek* offers, and second, to connect those lessons to contemporary science, particularly neuroscience and psychology. It may seem odd to think of an entertainment franchise, no matter how successful and long-running, as a significant source of insight and education. The *Star Trek* movies and television series are certainly not textbooks of psychology or neuroscience. Still, Gene Roddenberry knew what he was doing. His brainchild's characters and episodes accomplish things that textbooks can only dream about: they lead people to ask important questions while at the same time provoking wonder, curiosity, and learning.

Our aim is to use *Star Trek* as a vehicle for exploring inner space. This book will lead readers—confirmed, longtime *Star Trek* enthusiasts and neophytes alike—through some of the questions that *Star Trek* raises about mind and brain. Each of the following chapters runs a slalom course, weaving back and forth between *Star Trek* on one side and contemporary understanding of mind and brain on the other. In selecting topics for coverage, we subscribe wholeheartedly to the sentiments expressed by Leonard McCoy, M.D., ship's doctor on the U.S.S. *Enterprise*:

> *Blast medicine anyway! We've learned to tie into every organ in the human body but one: The brain! The brain is what life is all about.*
>
> (TOS-1, "The Menagerie")

For millions of people all over this planet in Sector 001 of the Milky Way galaxy, Kirk and Worf and Spock are mythic members of the family—the relatives whose exploits and foibles are gossiped about at the dinner table or at family reunions. Maybe you've puzzled over the odd behavior of your favorite aunt or the fascinating life your long-lost cousin has been living. Your relatives' idiosyncrasies can help you understand some of your own. You are family, after all.

Could it be that you've wondered about those even more distant relatives of yours, the colorful people you know only from *Star Trek*? Possibly, you've puzzled over

- why Spock can laugh but Data cannot
- why Nog's hearing is so amazingly good
- how Worf, a Klingon, is able to mate with Jadzia Dax, a Trill
- how Geordi LaForge, who's been blind since birth, can see better than sighted people can
- why the U.S.S. *Voyager's* Emergency Medical Hologram lost his memory
- how the entire crew of Deep Space Nine lost the ability to speak

The answers will come at the family reunion that we've arranged for you in these pages. Our hope is that by getting to know some of your distant—very distant—relatives, you'll have a better understanding not only of them, but of yourself as well. We are, after all, family.

Explanation of the Blurred Pictures

The two faceogenic pictures you saw earlier began life as the two completely random "checkerboard" patterns shown below. People don't see faces in these originals. After a computer has blurred each pattern, however, the lack of sharp edges allows the brain to interpret some of it as a face. If you are lucky enough to be extremely nearsighted, you may be able to see faces in the original patterns: just blur them by taking off your glasses or removing your contact lenses.[18]

Why Spock Can Cry but Data Cannot

> **DATA:** Mr. Comic, I wish to know what is funny.
> **HOLODECK COMIC:** Whatever makes you laugh is funny.
> **DATA:** Nothing makes me laugh. I wish to learn.
>
> (TNG-2, "The Outrageous Okona")

M r. Spock, the cool, calculating first officer of the first U.S.S. *Enterprise*, deplores emotions. To him, they are archaic nuisances, mere obstructions to good, sound reasoning. Spock prides himself on his relentless Vulcan-style logic and devotion to facts. But there are occasional cracks in his logical facade—those archaic nuisances sometimes get the best of him. For example, logical or not, Spock can cry. In the first of the *Star Trek* movies (*Star Trek: The Motion Picture*), we see Spock on the verge of tears. But usually he manages to keep a tight lid on his emotions. As he explains to Dr. McCoy, that's the Vulcan way.

> *I have a human half, you see, as well as an alien half.*
> *Submerged, constantly at war with each other. Personal*
> *experience, Doctor. I survive it because my intelligence*
> *wins out over both, makes them live together.*
>
> (TOS-1, "The Enemy Within")

Data, the android lieutenant commander aboard the *Enter-prise*-D, is a pale twenty-fourth-century version of Spock. Although

Data shares Spock's lack of emotional expression, he does not belittle emotions. Actually, he's fascinated by them. Watching his crewmates experience these strange unseen forces, Data craves a taste of what he's missing. To understand what it means to be human—Data's continuing mission—he yearns to experience emotions. But he cannot, because his constitution is different from Spock's, as it is from ours.

Spock is a biological creature—half-Vulcan and half-human, to be specific. He has a standard-model biological brain, as we learn when it is removed and put in charge of the operations of a small underground city (TOS-3, "Spock's Brain"). Data is different. He's an ultrasophisticated machine housed in a humanlike body. His brain consists of artificial neural circuitry that operates on the decay of positron particles. This positronic brain, as it's called, is a remarkable device. It can analyze vast quantities of data in the blink of an eye, and it can store more information than the Library of Congress. But it lacks the circuitry to produce emotions and feelings. Data cannot experience pleasure, anger, fear, or any of the other emotions that give color and definition to our lives. For all his positronic intellect, Data is stumped by a simple question: What does it mean to experience an emotion?

Angry Enough to Kill

Data and Lieutenant Commander Geordi LaForge are discussing Data's recent threatening encounter with the Borg (TNG-6, "Descent," I):

> DATA: Geordi, I believe that I've experienced my first emotion.
>
> LAFORGE: No offense, Data, but how would you know a flash of anger from some kind of a power surge?
>
> DATA: You are correct in that I have no frame of reference to confirm my hypothesis. In fact, I am unable to provide a verbal description of the experience. Perhaps you can describe how it feels to be angry. I could then use that as a frame of reference.
>
> LAFORGE: Yeah. OK. Well, when I feel angry first I . . . first I feel hostile.
>
> DATA: Could you describe feeling hostile?
>
> LAFORGE: Yeah. It's like feeling . . . belligerent. Combative.

DATA: Could you describe feeling angry without referring to other feelings?

LAFORGE: Hmm. No, guess I can't. I just feel . . . angry.

DATA: That was my experience as well. I simply . . . felt angry.

This exchange between LaForge and Data reminds us of something fundamental: emotional experiences are subjective mental states. Emotions are the exclusive property of the individual who has them. You can be sad because someone you care about is sad, but no matter how caring a person you are, you cannot really feel another person's pain. Nor can you experience their joy. The best you can do is to imagine how you might feel if you were placed in that individual's situation. You can never know for sure if you're feeling what your friend is feeling.

Like LaForge, you can try to find words to describe your feelings. But even if you succeed, those words remain just words, not feelings. And your words, however eloquent and precise, are useless unless you and your listener share a common frame of reference for interpreting them. That's Data's problem when he prods LaForge to describe anger without using words that refer to other feelings. LaForge can't do it, and neither could you.

Despite the communication gulf separating him and his close friend Geordi LaForge, Data is sure he has experienced genuine anger. After all, he "felt" something he'd never felt before, and it squares with what LaForge says *he* feels when he's angry. But must we take Data at his word that he's finally experienced his first real emotion?

Look at what happens just before Data's alleged flash of emotion. The android and three other members of the *Enterprise*-D crew beam aboard an alien ship, where they stumble upon a collection of mangled bodies. During their search for survivors, Data's tiny band runs smack into a squad of Borg (cybernetically enhanced creatures). An exchange of weapons fire kills one of Data's Starfleet colleagues. That's bad enough, but then one of the Borg grabs Data around the neck and begins choking him. This triggers something completely new in Data: his normally impassive demeanor gives way to what looks like full-blown rage. As he struggles with his Borg attacker, Data repeatedly warns him to let go of his throat. With each warning, Data's voice grows louder. He bares his clenched teeth in a threatening grimace, and his normally wide, innocent eyes narrow—reactions we humans associate with anger.

The Borg is no match for the android's superior strength. Data hurls the Borg against the ship's wall and then stands triumphantly over his battered foe. If you didn't know he was an android and that he had no emotion circuitry in his body, you'd swear that Data's facial expressions reflect an experience of intense emotion.[1]

Data's reactions make perfect sense under these circumstances. They look just like the natural, reflexive reactions you might have when confronted with a threatening situation. These strong affective reactions rivet your attention on the event that triggered the emotion, and mobilize your body for action. In response to a threat, emotion puts your body on **RED ALERT** to prepare you for physical confrontation. Your cardiovascular system routes blood to the major muscles, and your endocrine system pumps hormones, including adrenaline, into the blood. These physiological responses are so strong that you can feel them while they're occurring. The eminent American psychologist William James recognized the significance of this simple fact a hundred years ago. Turning conventional wisdom on its head, James proposed that *experiencing* these physiological reactions *caused* emotions, the reverse of the usual view. According to James, we don't blush because we're embarrassed; we're embarrassed because we find ourselves blushing. Whether James is right or not, his idea reminds us that part of the value of emotions lies in their ability to mobilize us for action.

The Universal Messengers

Strong emotions do more than get you ready to deal with a situation. Their visible and audible expressions send a message to others. When you're scared, blood vessels in the skin of the face constrict, reducing the volume of blood at the skin's surface. In Caucasians, this reaction produces a noticeable paling of the skin— we say that a frightened individual is as white as a ghost, or in an extreme case, as pale as Lieutenant Commander Data. When you're angry, just the opposite occurs. The blood vessels dilate, increasing the flow of blood to the facial skin and producing a flushed appearance. Face flush is particularly noticeable in fair-skinned individuals, such as Chief Engineer Miles O'Brien of Deep Space Nine. The Ferengi Quark teases O'Brien by noting that the Chief's face gets really pink whenever he's aggravated (DS9-1, "Captive Pursuit").

Quark is a keen observer, but he's no match for Constable Odo, the shape-shifter who's in charge of security on Deep Space

Nine. Among Odo's sleuthing skills is the ability to detect when someone is lying or withholding crucial information. And among the cues he uses for spotting deception are the facial expressions of the person he's interrogating. During the Cardassian occupation, at which time Deep Space Nine is a mining station known by the Cardassian name Terok Nor, Odo has to find the person who murdered a Bajoran chemist (DS9-2, "Necessary Evil"). Odo's interrogations include a young Bajoran woman who has recently come aboard the mining station. The woman is Kira Nerys, then a member of the Bajoran underground, but later second-in-command of Deep Space Nine. Odo is suspicious of Kira because she's already lied about where she was on the night of the murder. He confronts her about her false alibi:

> **KIRA:** Yes, I lied about my alibi. That doesn't make me a killer.
>
> **ODO:** Where were you when he was murdered?
>
> **KIRA:** Asleep, alone.
>
> **ODO:** No one saw you in the community quarters?
>
> **KIRA:** I wasn't there. I found a small corner and . . .
>
> **ODO:** You're lying.
>
> **KIRA:** I . . .
>
> **ODO:** Don't bother. Your whole face changes. Should have seen it before. You don't lie well.

Odo may have a gift for reading faces, but he's not alone. For centuries, people have exploited facial expressions to gain access to other peoples' feelings and intentions. The Greek philosopher Aristotle knew the value of reading faces: "There are characteristic facial expressions which are observed to accompany anger, fear, erotic excitement, and all the other passions."[2] Centuries later, Charles Darwin called attention to the link between emotion and facial expression; in fact, he made it a linchpin in his arguments for the continuity of animals and man. In the facial expressions of dogs and monkeys, Darwin believed he saw precursors of human expressions of emotion.

Today, psychologists usually speak of six basic emotions, each associated with a particular facial configuration. As you can see in the photographs on the following page, these emotions are distinguished by the shape of the mouth, the width of the eyelids, and the disposition of the eyebrows.[3] All of these emotional signals are produced by contractions of the musculature of the mouth and

Happiness

Surprise

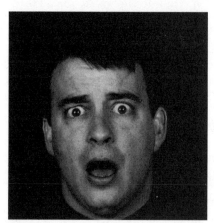

Fear

Sadness

Anger

Disgust

eyes. These subtle and not-so-subtle signals make it possible for other people to know what you're feeling. Animals, including cats and dogs, also use their ears to signal when they're feeling angry (ears back close to the head) or friendly (ears forward and perked up). Humans get by on just eyes, eyebrows, and mouths.

Darwin believed that these complex patterns of muscle contraction occur instinctively—that learning them is unnecessary. Any parent can attest that a newborn infant exhibits a wide range of facial expressions, and that those expressions are correlated with the infant's emotional state. For example, an infant with a soaking wet, cold diaper is unlikely to be cooing and smiling; ear-shattering cries and copious tears are far more likely—and far more effective in getting the problem taken care of. A human infant's precocious emotional expressiveness tells us that the brain's control over our facial muscles is present at birth, an innate part of our "nature."

Comparisons of emotions across different cultures point to the same conclusion: nature plays a strong role in shaping our emotional lives.[4] The best of these cross-cultural comparisons didn't wait around for emotions to happen; instead, researchers intentionally provoked emotions, aided by cinematic gore—remember that great bloody stabbing-in-the-shower scene in Hitchcock's *Psycho*? While volunteers watched gory scenes from movies, the researchers measured physiological indicators of their emotions: heart rate, blood pressure, and the electrical conductance of the skin. Each of these measures is an index to the amount of activity within a person's autonomic nervous system, that subdivision of the nervous system whose activity mirrors our emotional state.

This experiment was done first with volunteers recruited from a university in the United States. As they watched different movie scenes, the volunteers showed emotion-specific patterns of autonomic nervous system activity. These patterns were so specific that a researcher could look at the record of a volunteer's physiological measurements and tell quite accurately what emotion—fear, anger, disgust, or sadness—was being provoked. When the same experiment was repeated with members of the Minangkabau society in western Sumatra, they showed the same patterns of emotion-specific activity that had been seen in American college students. This commonality of physiological responses points to some underlying biology that is universal and innate—just as Darwin thought.

Rules of Emotional Expression

Darwin may be right about the biological origins of emotions, but that doesn't mean that learning and culture play no role in their expression. Experience teaches each of us when it's appropriate to express emotions and what forms those expressions can take. As we're growing up, we learn the rules for emotional expression from our parents and others around us. The android Data, who has no emotions, has not been schooled in those rules, and that's where he gets into trouble.

In the feature-length movie *Star Trek VII: Star Trek Generations*, Data becomes increasingly discouraged by his failed attempts to understand emotions. He notices that many humans take pleasure in playing practical jokes on one another. In the timeless spirit of "android see, android do," Data plays what he believes will be a terrific joke on Beverly Crusher: he shoves the fully clothed doctor into cold water.

Data's closest friend, Geordi LaForge, scolds him for his dumb joke. Worried about the consequences of future inappropriate actions, Data decides to have an emotion chip installed in his positronic brain. Data's creator, Dr. Noonien Soong, who invented the chip, had meant to put it into Data's brain one day, but never got around to it, so Data has been saving the tiny electronic circuit for a rainy day.[5] Data decides that rainy day has come, and solicits LaForge's help with the chip's installation.

The initial results are not what either Data or LaForge expects. Data spews out a stream of puns, sings silly songs, and laughs uncontrollably as he mentally replays jokes he heard years ago. He didn't get them then, but thanks to the chip, he gets them now.

DATA: I get it! I get it!

LAFORGE: You get what?

DATA: When you said to Commander Riker "The clown can stay but the Ferengi in the gorilla suit has to go."

LAFORGE: What are you talking about?

DATA: During the Farpoint mission. We were on the bridge. You told a joke. That was the punchline.

LAFORGE: Data, that was seven years ago.

DATA: I know. I just got it. Very funny.

Data's newly installed chip also introduces him to the darker side of emotions. When he and LaForge are on an away mission, they are attacked by a deranged doctor who is in cahoots with Starfleet's enemies. The doctor knocks LaForge unconscious and then threatens to kill Data. With the doctor's phaser staring him in the face, Data, for the first time in his thirty-four-year existence, experiences paralyzing fear. As he cowers in a corner, he begs for his life. Even after the immediate threat to his life passes, Data's terror is still so overwhelming that he can't mobilize himself to help LaForge.

We learn two simple but important lessons from Data's behavior following his chip installation and sudden infusion of emotions. First, while there are times when it's appropriate to express an emotion, there are also times when it's definitely inappropriate. You can laugh when a friend accidentally spills beer on his blue jeans, for example, but not when your father-in-law spills red wine on his new white linen sports coat. Sharing a risqué pun with a close friend may be harmless, but repeating the same pun in front of your Sunday school class may lead to rebuke or even threats of divine retribution. It's OK to cry in front of your spouse, but not in front of your boss. (If your spouse *is* your boss, all bets are off.) Data is an emotional novice and hasn't had enough experience to learn this lesson. His nonstop attempts to be funny become insufferable to those around him. It takes time, but Data finally learns the ancient human art of restraint.

Data also has a thing or two to learn about the emotion we call fear. Fear is a natural response to danger. In fact, fear is often a good thing, because it can mobilize the body for appropriate action. But the very first time Data experiences fear, it stuns and paralyzes him. It doesn't take any practice to experience fear, but it may take some time to get the hang of what to do with it. Police officers and firefighters are living proof that people can learn to cope with situations that might otherwise evoke paralyzing levels of fear.[6]

In parts of the United States today, many children's "free" hours are gobbled up by lessons of all sorts: dance, piano, manners, computer, swimming. But not even the most doting parents enroll their children in regular emotional expression lessons. Still, in every culture, children receive countless hours of tutoring in emotion's unwritten rules. Their tutors are family members, peers, mass media, and just about everyone else with whom children come in contact.

Unlike the rules for international games such as chess and tennis, the rules for emotional expression vary from culture to culture. And this variation can cause misunderstandings. In one study of emotional expression, groups of volunteers were asked to view films depicting unpleasant scenes, such as the amputation of an arm. One group was made up of Americans, the other of Japanese. As the groups watched the films, they were joined by researchers who watched their reactions—which were amazingly different. The American viewers exhibited strong negative reactions, including disgust, while the reactions of the Japanese viewers suggested placidity and, at times, even amusement.

To American eyes, the responses of the Japanese may seem strange. But these peculiar responses do not prove that their emotional reactions differed from those of the Americans. Instead, their Spock-like impassivity reflected a learned cultural taboo against showing negative emotions in the presence of authority figures or strangers. This became clear when the experiment was repeated with one small change: the Japanese viewers were led to believe that they were watching the films in private, with no authority figures to monitor their reactions. As a result, they showed strong negative reactions that were indistinguishable from those of the Americans.

We often see such cultural differences played out in sports and recreational games. In the United States, National Football League players who celebrate too much after a score by their team can be penalized for unsportsmanlike conduct. And this rule is tough: a player can be fined merely for removing his helmet. Things are different in the Southern Hemisphere: Latin American soccer players behave deliriously when a goal is scored. Their exuberant celebration is not penalized; in fact, jumping around, hugging, and shouting are expected and encouraged.

Poker, a favorite pastime among the officers of the *Enterprise-D*, is another international game in which emotional expression is important. A successful poker player learns to mask her elation when she draws an ace-high straight, lest other players sense her excitement and fold without raising the ante. And when she draws a losing six to the potential straight, she must manage to hide her disappointment, in hopes of bluffing her way into a win. Among Starfleet personnel, Commander William Riker is famed for his expressionless poker face. His bluffs at the poker table usually pay off (TNG-6, "Second Chances"; TNG-7, "Lower Decks"), but other times they cost him plenty (TNG-3, "Best of Both Worlds," I). Unlike Riker, Benjamin Sisko, captain of Deep Space Nine, has never mastered the art of bluffing. Starfleet officers travel from the farthest reaches of the universe for a chance to get Captain Sisko to the card table. As Sisko's science officer explains to his second-in-command, "I've worked for two lifetimes on Benjamin's poker. He just can't learn how to bluff" (DS9-2, "Paradise").

What Do You Expect?

Two Starfleet officers aboard Deep Space Nine take some time off to play a friendly game of chess. But things quickly deteriorate. One accuses the other of cheating, knocks the pieces off the chessboard, and stomps out of the room. Bad sportsmanship? For sure. But what else can we say about this bad sport? Before you can answer, you have to know who it is. If the poor loser were Chief O'Brien, the mini-tantrum might be chalked up to his Celtic temper. If it were Worf, we might blame his Klingon competitiveness. But the chess players are actually Benjamin Sisko and his long-time friend Jadzia Dax (DS9-3, "Equilibrium"). And surprisingly, it's the normally controlled Dax who loses her cool. This outburst is one of several atypical eruptions of anger that result from an abnormally low level of neurotransmitter in her nervous system. Because such behavior is so unusual for her, other crew members immediately know that something is wrong.[7]

Dax's outburst, and the onlookers' reactions to it, are worth thinking about. What you make of someone else's emotional state depends on that person's characteristic baseline behavior pattern. Such a pattern is called "temperament," and it is an enduring personality trait. On space station Deep Space Nine, Benjamin Sisko is typically forthright, Odo impassive, Miles O'Brien irascible, Julian Bashir high-spirited, and Quark, well, devious. These adjectives describe the characters' temperaments. Yet, just as there is an occasional warm afternoon in winter, people can and do deviate from their temperamental baseline. And it is such deviations that constitute emotional reactions.

Some people's temperaments are so extreme that it's difficult to gauge their emotions. For example, Dr. McCoy of the original *Enterprise* is nearly always agitated, irritated, and popping off about one thing or another, so it's hard to know when something serious is troubling him. In contrast, think back to the time before Data receives his emotion chip, when he reacts so violently upon being attacked by the Borg. The extent to which he deviates from his normal calm demeanor makes it obvious that he is experiencing something out of the ordinary. Data has never exhibited such an intense reaction before, even though he has been in plenty of dangerous, threatening situations. He remains dispassionate and calm even when informed of the death of his daughter Lal (TNG-3, "The Offspring"). Against this background of steady unflappability, his violent outburst suggests that he is indeed experiencing an emotion, his very first.

In humans, temperament depends on brain chemistry. Some people—optimists, we call them—are just naturally happy, always managing to see the bright side of things. And their optimism appears to be an enduring state that tends to survive temporary setbacks and negative life events, such as the loss of a job. At the opposite end of the spectrum are those individuals who seem chronically pessimistic, expect the worst, and often get it. Compared with their optimistic counterparts, chronically negative individuals are more prone to depression and drug addiction. And despite what we might think, positive life events, such as winning the lottery, don't transform pessimists into optimists.[8] Instead, their normally dour dispositions endure. Temperament, in other words, is an enduring personality characteristic.

Your heredity influences your temperament; happiness (and its opposite) is grounded in biology. We know this from studies that use paper-and-pencil tests to characterize a person's general emotional outlook. In such studies, test takers are asked to indicate how much they agree or disagree with statements such as "I am just naturally cheerful" and "My future looks very bright to me." From the answers, researchers can calculate an overall score that reflects how positive the test-taker's emotional outlook is. When identical twins take this kind of test, their scores tend to fall in the same range—that is, you can predict one twin's score based on knowledge of the other's. This similarity holds even if the twins were reared apart. It does not, however, hold in the case of fraternal twins, in which the brightness of one sibling's outlook is unrelated to that of the other.[9] Identical twins, of course, are genetically identical, whereas fraternal twins are not.[10] So the similarity between identical twins' test scores, particularly in the case of twins reared apart, points to a hereditary basis for emotional outlook.

Genes influence a person's outlook by way of a specific neurotransmitter system within the brain: the dopamine system. The neurotransmitter dopamine is manufactured by neurons in the brain stem, near the base of the brain. Their axons extend upward from these dopamine-manufacturing cells and fan out to make contact with dopamine-sensitive neurons throughout the brain. Neurons that respond to dopamine are particularly numerous in the limbic system and the prefrontal cortex. These two regions of the brain are major players in the control of our emotional reactions, which explains how dopamine gets into the "happiness game."

Dopamine is the brain's "pleasure chemical." This may sound like an advertising slogan, but it's true. Dopamine floods the brain whenever someone experiences pleasure, regardless of whether that pleasure springs from a good meal or a snort of cocaine. The differences in people's temperaments are probably related to their sensitivity to dopamine. For one thing, we know that people vary widely in their responsiveness to drugs that mimic dopamine. Individuals who score high on the positive emotional outlook scale tend to respond strongly to such drugs. We can't be certain whether these chronically "up" individuals have more dopamine receptors in their brains, or whether the receptors they have are just more sensitive. Either way, their temperaments reflect a structural property of their brain cells. Thanks to dopamine, these individuals do have more fun.[11]

Dopamine is obviously a good thing to have. But like most good things, one can have too much of it. An extra-strong dose of dopamine, or a highly exaggerated response to a normal dose, can push a mind over the edge. While there are clearly other factors involved, we know that dopamine plays a role in schizophrenia. We don't know for sure whether the schizophrenic brain manufactures too much dopamine, or whether its neurons are just hypersensitive to dopamine's effects. Either way, the key problems brought on by schizophrenia—disturbed thought and blunted affect—match up with the functions of those regions of the brain whose neurons are sensitive to dopamine: the prefrontal cortex (thought processes) and the limbic system (emotions).

Serotonin is another neurotransmitter that's important in regulating temperament. Like the dopamine-manufacturing cells, the neurons that make serotonin are located in the brain stem, and they, too, send their axons to the limbic system and throughout the cortex. Abnormally low levels of serotonin are associated with several mood disorders, including depression and chronic anxiety. Drugs like Prozac, which is widely prescribed for depression, do their job by extending the time that serotonin molecules remain in the synapses between cells, which in turn increases the time that these molecules have to stimulate serotonin-sensitive neurons. As a result, the neurotransmitter molecules have a stronger effect.

When Expressions of Emotion Are Deceiving

When circumstances dictate, the expression of emotions can be muted, repressed, or even momentarily distorted. For example, almost every photograph taken of England's Queen Victoria after 1861 shows Her Majesty glowering at the camera. From the large number of such photographs, you'd think that she was stern and perpetually glum, perhaps mourning the death of Albert, her beloved consort. But this apparent glumness probably reflected the Queen's dislike of photographers rather than her temperament.[12]

Queen Victoria's present-day counterparts have also been accused of being "coldhearted and uncaring." Along with other less flattering terms, that is how many British citizens and the news media characterized the Royal Family's lack of emotional response to the death of Princess Diana in 1997. But let's be honest with ourselves—no one can be certain what feelings the Royal Family experienced or didn't experience. No matter how cunning or obtrusive the paparazzi may be, their cameras cannot peer into the brain.

The lesson? When it comes to emotions, you can't judge a book by its cover. An expression of emotion—particularly a single, momentary expression—can be a very unreliable clue to someone else's feelings. Even experts sometimes make incorrect inferences about the emotional states of others. Consider an experiment in which student nurses were asked to describe a series of cheery, delightful films they were watching.[13] This may sound easy, but there was a catch: only half of the films actually fit that description. The others were anything but, containing images of people with severe burns or undergoing limb amputations. Regardless of the film they were watching, the nurses were instructed to convey the impression that it was a pleasant film, even if that meant overriding their real feelings. To get a feel for how difficult this task must have been, imagine watching a gruesome scene from *The Texas Chain Saw Massacre* or *Friday the 13th* and trying to convince someone you were watching "The Trouble with Tribbles" (TOS-2).

To make things even more challenging for the nurses, they had to describe the films to FBI agents, psychiatrists, police officers, judges, and other trained professional lie spotters. These experts listened to the nurses' descriptions and judged whether or not they were telling the truth. The outcome was rather embarrassing to the supposed experts: these vaunted lie detectors succeeded only 50 percent of the time in identifying when the nurses were lying. Some nurses were particularly good at fooling the professionals; others were less able liars.[14] Maybe Garak, the Cardassian tailor on board Deep Space Nine, is correct when he boasts that lying (which he is very good at) is a skill that, like anything else, must be practiced if one wants to do it well (DS9-5, "In Purgatory's Shadow").

How do people like these nurses manage to mute their expressions of anger or disgust under highly charged emotional circumstances? And how does Commander Riker disguise his emotions when he's playing poker with his Starfleet cronies? The answers to these questions lie deep within the recesses of Spock's brain.

A Tour of Spock's Brain

When Spock learns that a close colleague has just been murdered, this terrible news elicits absolutely no emotional reaction from the stoic Vulcan (TOS-1, "The Man Trap").[15] Lieutenant Uhura, a human *Enterprise* crew member, is aghast at this, and angrily denounces Spock for his lack of compassion (much like the public criticized the British Royal Family after Princess Diana's death). In his customary composed, rational manner, Spock reminds the lieutenant that compassion will not change the outcome of this unfortunate turn of events.

We know that Spock can experience emotions. We have seen him experience sadness (*Star Trek: The Motion Picture*), amusement (TOS-3, "Plato's Stepchildren"), and anger (TOS-1, "This Side of Paradise"; TOS-3, "Day of the Dove"). Therefore, his lack of emotional display upon learning of his colleague's death means that Spock is merely keeping a lid on the emotions he so clearly possesses. But how?

For many years scientists searched for a single "center" within the brain that was responsible for all emotional reactions. The original focus of their search was the limbic system,[16] a confederation of neural structures located deep within the brain. It was a reasonable starting point for their search, since they knew that if an experimenter surgically destroys an animal's limbic system, the animal seems to lose its sense of fear, and no longer distinguishes dangerous situations from safe ones. A surgically altered rat, for example, might stroll right up to a sleeping cat and nibble casually on its ear.

But subsequent evidence forced researchers to abandon the notion of a single emotional center in the brain. Nowadays emotions are seen as the product of neural activity that is distributed over many brain regions. Although the limbic system is one of those regions, other structures are equally involved. One of these structures, the prefrontal cortex, deserves special attention, for it may hold the key to understanding Mr. Spock and others like him.

Starfleet medical personnel use a simple lightweight device to peer into various parts of a patient's body, including the brain (VOY-4, "Scorpion," II). This amazing device, the medical tricorder, allows Doctors Bashir and Crusher, as well as the Emergency Medical Hologram, to see which parts of a patient's brain are more active than normal, which are less active, and which are "just right."[17] Twentieth-century medical technology has not yet produced an easy-to-use handheld instrument that performs the functions of the medical tricorder. But it has given us large, complex neural imaging devices that can give us a rough idea of which brain areas are active and which are quiescent. We don't know how the medical tricorder works, but the principles behind its clumsy twentieth-century counterparts are simple. When brain cells are active, their metabolism increases. To fuel this increased metabolic demand, the blood must deliver extra glucose and oxygen to the active region. Various technical tricks can convert these delivery patterns into pictures of the brain at work. One such technique, positron emission tomography (PET), requires the injection of a minute amount of radioactively labeled water into the patient's bloodstream. Sensors placed around the skull identify those regions of the brain in which the radioactivity is most highly concentrated. Another technique, functional magnetic resonance imaging (fMRI), is even simpler and does not involve radioactive labeling. Instead, sensors detect increases in the blood's oxygen concentration in local areas of the brain.

Neural imaging makes it possible to view the anatomical distribution of brain activity in people experiencing heightened emotions. The resulting images reveal a wide distribution of active brain areas. As expected, these areas include parts of the limbic system. However, portions of the cerebral cortex at the very front of the brain are also strongly activated. Take a look at the photograph below—it's a PET image of the brain of a person who is emotionally aroused. We're viewing the brain from above, with the activated regions shown in light gray. These regions include the frontal lobes, and the area at the top of the picture corresponds to the brain's prefrontal cortex. Like many other anatomical terms, the name "prefrontal cortex" specifies the region's location in the brain: it occupies the most forward position in the brain, just above the eye sockets.

Within the nervous system, as in real estate, location is everything. And nowhere is this truer than in the prefrontal cortex. The neurons of this brain area are ideally located to coordinate three types of information that are crucial when it comes to emotions: sensory (information about events in the external world), memory (information about the relevance and significance of those events), and internal (information about the organism's

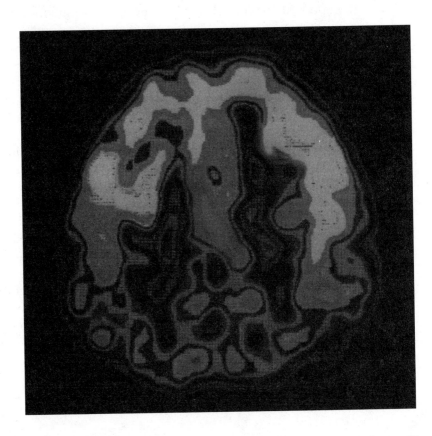

physiological needs). Of course, emotions are most useful if they lead to some action (hopefully, an appropriate one), and the prefrontal cortex is well positioned to contribute to such action. Its location allows for extensive connections to the portions of the brain that initiate voluntary motor activities, such as playing the banjo, tying your shoe, or—most relevant here—slamming your fist against a wall in anger. Thus, three types of information converge on a hub of neurons that has special control over voluntary actions. A near-fatal transporter accident shows how all these parts come together when emotion is involved.

The Emotional Side of Rationality

In the twenty-third century, transporters are still a developing technology. Transporters convert a person's matter into an energy profile and beam that energy to the desired remote location, where it is transformed back into the person. At least that's how it's supposed to work. Like any new technology, transporters sometimes malfunction. These malfunctions can take various forms. Starting with the least dangerous, sometimes a transporter can't lock onto the person who is to be transported. Like an airline flight that's canceled, the transport is aborted and the person stays in place. No harm done. Occasionally, the person is transported all right, but, like a passenger on a flight diverted from its destination, ends up in the wrong place.[18] Annoying, but nothing that can't be fixed. Then there are those catastrophic failures in which the person ends up in the correct spot, but fails to rematerialize properly. For this sort of transporter error there is no twentieth-century analogue.

One of these freak errors splits Captain Kirk into two persons, whom we'll call Kirk I and Kirk II (TOS-1, "The Enemy Within"). While the two Kirks look identical,[19] their temperaments couldn't be more different. Kirk I is emotional, rash, and dangerously impulsive; he sexually attacks Yeoman Janice Rand, and clobbers Geological Technician Fisher. Kirk II is meek and indecisive. His emotions are blunted, and he can't make decisions about anything—a deadly trait for a starship commander. Paralyzed by his indecisiveness, Kirk II sits placidly on board the *Enterprise* while some of the ship's crew members are freezing to death on the planet below—he is reluctant to beam them up, because another transporter accident could split them, too.

Spock, ever the rationalist, recognizes the implications of Kirk's predicament and shares his insight with Dr. McCoy:

SPOCK: We have here an unusual opportunity to appraise the human mind. Or to examine, in Earth terms, the roles of good and evil in a man. His negative side, which you call hostility, lust, violence. And his positive side, which Earth people express as compassion, love, tenderness.

McCOY: It's the Captain's guts you're analyzing. Are you aware of that, Spock?

SPOCK: Yes. And what is it that makes one man an exceptional leader? We see here indications that it is his negative side which makes him strong. That his evil side, if you will, properly controlled and disciplined, is vital to his strength. . . .

Fortunately for Kirk—and for his freezing crew members—the malfunctioning transporter is repaired and his two embodiments are reunited, restoring his ability to make decisions and command his ship. But what happened during the split? Why did Kirk II's loss of emotional tone knock out his ability to make decisions? What's the connection between the two?

In the Division of Cognitive Neuroscience at the University of Iowa, Antonio Damasio studies twentieth-century humans who bear an eerie resemblance to Kirk II. These patients have suffered damage to the prefrontal cortex of the brain. Whether from tumor or stroke, this damage leaves the patients with their intelligence intact, but with a couple of other notable deficits. First, these patients typically exhibit little emotional expression. They always appear cool and detached. According to Damasio, these people seem to know, but not to feel. Listen to Damasio describing the blunted affect of a thirty-five-year-old accountant, code-named E, who had undergone surgery to remove a tumor that had invaded his prefrontal cortex:

> E was able to recount the tragedy of his life with a detachment that was out of step with the magnitude of the events. He was always controlled, always describing scenes as a dispassionate, uninvolved spectator. Nowhere was there a sense of his own suffering. . . . E was exerting no restraint whatsoever on his affect. He was calm. He was relaxed. . . . He was not inhibiting the expression of internal emotional resonance or hushing inner turmoil. He simply did not have any turmoil to hush.[20]

Prefrontal damage also creates another deficit, less obvious but more troubling than blunted emotions: patients experience agonizing difficulty with decisions. They can take hours making simple decisions that we make with barely a moment's thought. Here's Damasio's description of E's decision-making behavior:

> He needed about 2 hours to get ready for work in the morning, and some days were consumed entirely by shaving and hair washing. Deciding where to dine might take hours, as he discussed each restaurant's seating plan, particulars of menu, atmosphere, and management. He would drive to each restaurant to see how busy it was, but even then he could not finally decide which to choose. Purchasing small items required in-depth consideration of brands, prices, and the best method of purchase. He clung to outdated and useless possessions, refusing to part with dead houseplants, old phone books, six broken fans, five broken television sets, three bags of empty orange juice concentrate cans, 15 cigarette lighters, and countless stacks of old newspapers.[21]

Patient E couldn't make up his mind to jettison the junk. He labored long and hard over the most trivial matters. But, oddly, when dealing with important decisions, he could be downright reckless. E ignored risks in financial investments and business enterprises. Following his brain surgery, he ran through his life savings and, ultimately, went bankrupt.

There's something very odd here. Sometimes E made rash decisions, and other times he couldn't make any decisions at all. Sometimes he was like Kirk I, other times like Kirk II. Ordinarily, we don't think of emotions and decision making as intimately related. Most of us are like Spock: we believe that decisions should be made dispassionately, with only the facts in mind. But patients like E show us that effective decision making goes hand-in-hand with emotions. Remove emotions from the mix, either by brain injury or transporter malfunction, and wise decision making is impossible.

Hidden Emotional Records

Jadzia Dax, science officer on board Deep Space Nine, is a Trill. These unusual aliens are a joined species consisting of a host (the visible body) and a symbiont (a slimy wormlike creature housed within the host's body). A symbiont cannot live on its own, so when its host dies, it must be quickly transplanted into another host. Because symbionts are much longer-lived than hosts, each symbiont typically inhabits a number of hosts during its lifetime. Memories accumulated during one host's lifetime are carried by the symbiont to its next host. This makes for complicated memories and heavy emotional baggage. Jadzia Dax, for example, describes to Benjamin Sisko the pain of watching a young child suffer, an emotion and a memory left over from an earlier host of her symbiont (DS9-5, "Nor the Battle to the Strong").

Humans are not so different from Trills. We retain memories of emotions, and those memories lend an affective overtone to our decision making. Damasio, the University of Iowa neurologist, has a name for these memories—*somatic markers*—and, more importantly, a theory about how they work. (The word *somatic* means "of the body," and *marker* refers to a tag or signpost.) According to Damasio, our emotionally charged memories of places, objects, and even events become marked. These markers carry a somatic signature, meaning that they evoke a bodily feeling. Once you've had a painful experience in the dentist's chair, you experience heightened nervousness each subsequent time you approach the dentist's office. When you've enjoyed a pleasant afternoon in the park with a friend, you feel warm and happy as you approach the park a week or a month later. When something in the environment activates a somatic marker, a version of the original feeling is automatically reinstated.

Somatic markers improve decision making by providing valuable clues about the consequences of alternative courses of action. Negative somatic markers are warning bells that reduce the attractiveness of particular alternatives. Positive somatic markers boost the attractiveness of alternatives that are so marked. If the brain regions responsible for planning are denied access to somatic markers, chronic indecision results. This is probably why Kirk II and Damasio's patient E behave as they do.

Negative markers, those emotional warning bells that can be so useful, are generated in the amygdala, a small almond-shaped cluster of neurons that is part of the limbic system. If a neurosurgeon stimulates the amygdala in an awake human patient whose brain has been exposed for some surgical procedure, the patient reports a flash of fear or anxiety.

The amygdala enjoys priority delivery of sensory input. Neural signals about events in the environment reach the amygdala before more refined versions of those signals reach the cortex and break into awareness. Because of this express delivery, the amygdala can generate responses from a quick, but unrefined, analysis of a situation. This puts you in a strange position: you can respond to potentially dangerous events before you actually recognize what those events are. Imagine walking through tall grass when out of the corner of your eye you catch a fleeting glimpse of something moving. If you've run into snakes in this vicinity before, there's a very good chance that you'll jump back several steps, even though at the moment of the jump you don't even know that you've seen anything. This reflexive, emotionally driven reaction occurs courtesy of your amygdala. Like any preliminary draft, the fast first draft written by the amygdala can stand some rewriting by the slower, more accurate cerebral cortex. Perhaps that movement in the grass was really a harmless foraging chipmunk. You jumped for no good reason—but better safe than sorry.

Emotion slaps a Day-Glo marker on some memories: **NOTE: THIS MEMORY IS REALLY IMPORTANT!** This marker gives a special privileged status to the memory it's attached to. Who was the first person with whom you shared a long, romantic kiss? Where did that kiss take place? Where were you when the space shuttle *Challenger* exploded? A dollop of emotional juice enhances a memory's vividness. Or, in Damasio's parlance, somatic markers highlight memories of special, emotionally significant events. Memories of these events usually come back very quickly and with unusual clarity. Speed and clarity, however, are not the same thing as accuracy. No matter how convincing an emotionally charged memory seems, it is still just a memory. The memory of your first real kiss may be correct, but its details are not immune to forgetting, confusion, or outright revision.[22]

Somatic marking is important, but its opposite—somatic *un*marking—is just as important. Removing or changing a memory's warning label is one way in which the brain constantly updates its information about the world. After Kirk had his unfortunate run-in with that malfunctioning transporter, his memory of the transporter was probably marked **DANGER! KEEP AWAY!** But once the transporter is fixed and is no longer dangerous, that label should come off. Conditions change, and our responses, emotional and otherwise, should change along with them.

Sadly, some people have trouble unlearning fear responses that may have been appropriate once, but are now out of date. In post-traumatic stress disorder, a traumatizing event alters the brain so that years later, a similar but harmless event triggers a flashback, in which the original emotions are reexperienced. A traffic helicopter flying overhead, for example, can transport a traumatized combat veteran back to a battle that occurred decades ago.

How Spock Keeps His Cool

By human standards, Spock's self-control is remarkable. How does he remain impassive when all the humans around him are losing their heads (figuratively speaking)? Because Spock prides himself on his Vulcan logic, let's approach this question logically. There are two possibilities.

The first possibility is that Spock's autonomic nervous system fails to generate the bodily reactions normally associated with emotions. If this were true, Spock wouldn't feel the physiological changes that accompany emotions. Therefore his brain would have no reason to tag his experiences with somatic markers. According to this hypothesis, Spock's mental records of past events and circumstances should be factually correct, but affectively neutral, unmarked by splashes of emotion. Consequently, Spock's body should never show telltale signs of emotion. When he faces a situation that would stress a human, he shouldn't sweat; when he's in a pinch that would embarrass a human, he shouldn't blush; and, in an encounter that would scare a human, he shouldn't get telltale goose bumps. A medical tricorder should show that Spock's heart rate, blood pressure, and skin conductance are rock steady no matter how threatening or exciting his situation.

This prediction does not stand up to the evidence—Spock's body does generate emotional reactions. On one occasion, Spock and some other *Enterprise* crew members are infected by spores from an alien plant (TOS-1, "This Side of Paradise"). The infection produces an overwhelming contentment, which turns every day into a day off, killing any desire to work. Captain Kirk, who is not infected by the spores, discovers that adrenaline readily destroys them. Kirk knows that the adrenal gland pumps out adrenaline when one is intensely aroused. He decides to use this knowledge to cure his first officer of his dangerous contentment. To get his adrenaline pumping, Kirk unmercifully ridicules Spock's Vulcan origins. The treatment works: Spock becomes enraged, pumping adrenaline into his bloodstream. Spock is restored to his usual well-controlled self.

This incident shows that Spock's body has the physiological machinery for creating emotional responses, which are key ingredients of somatic markers. So the first possible explanation for Spock's flat affect is certainly wrong—but there is another.

The second possibility is that Spock's body generates normal somatic markers, but his brain doesn't link them to information about the situations that evoked the markers. According to Damasio's scheme, this is precisely what is wrong with human patients with damage to the prefrontal cortex. To see whether this is Spock's situation as well, we can compare his behavior with that of Damasio's patients. In one respect, their behavior is identical. When Damasio's patients are confronted with disturbing scenes (for example, a picture of an automobile accident), they show absolutely no signs of arousal or distress. The patients acknowledge that the events in the pictures are grisly, but they are completely unaffected by what they're seeing. The same is true of Spock. Even when he is informed that his father, Sarek, has died, Spock is impassive (TNG-6, "Unification," II).

So far, the dispassionate Spock looks a lot like Damasio's dispassionate patients. But there is an important difference: the brain-damaged humans are unconcerned about their behavior's possible negative consequences. In high-risk situations, where normal humans are cautious, these patients are breathtakingly reckless. But that's not Spock. He seems constantly mindful of the potential consequences of his actions—good and bad. This implies that Spock's brain does lay down somatic markers that can shape his subsequent behavior.

Despite his reputation for bloodless rationality, and despite his lectures on the evils of emotions, Spock admits that he experiences emotions. When Dr. McCoy launches a tirade against Vulcans—remember, Spock is half Vulcan and is very proud of his Vulcan heritage—Spock tries to punch McCoy (TOS-3, "Day of the Dove"). He chalks this up to a temporary wave of racial hatred. Years later, Spock acknowledges his emotional side during a conversation with Captain Jean-Luc Picard (TNG-6, "Unification," II). Picard is on the planet Romulus to find out why Spock is making an unauthorized visit there. Spock explains that his mission had to be a secret or many lives would have been jeopardized. He confides in Picard that his decision was colored by emotions, and acknowledges that emotions can be useful. They have given him the ability to see beyond pure logic.

We are left with one conclusion: Spock's vaunted lack of emotion is more illusory than real. He *feels* emotions and makes good use of them, but he usually manages to suppress their expression.

In this sense, Spock resembles the student nurses who managed to hide their revulsion when they watched gruesome films, or the Japanese, who stifle displays of negative emotions in certain culturally sanctioned conditions. Or Commander Riker, who masks his elation when he draws a winning poker hand.

So what's going on here? Emotions are biological forces, with millions of years of evolution behind them. How can someone override something as powerful as an emotion? The phrase "mind over matter" captures what happens when you succeed in squelching the expression of some emotion. And there's no particular mystery about this process—for your brain, it's just part of the daily routine.

Among your brain's billions of neurons there are countless interconnections. Each individual neuron makes contact, on average, with thousands of others. The total number of contacts is staggering: in the neighborhood of 100 trillion. These interconnections allow the various regions of the brain to have ongoing conversations with one another. Through these conversations, ensembles of neurons in one part of the brain can influence the activity of neurons elsewhere. The results of these neural dialogues are the plans you make and the actions you take. These same dialogues make it possible to inhibit actions that might be inappropriate. The prefrontal cortex can edit or veto the amygdala's urgings, killing the expression of an emotion. (Think about Riker sitting at that poker table.) Here, "mind over matter" means that one set of brain regions exercises a veto over the commands of another.

Like Spock, we can learn to control many of the brain's functions that are normally considered to be automatic or reflexive. For example, simple biofeedback techniques let you influence your own autonomic nervous system, even to the extent of regulating your blood pressure and heart rate. Although we don't yet know the details of the neural pathways involved in control of emotional expression, we do know that whatever they are, Spock's brain has them. And fortunately, so does ours.

Data Learns a Lesson

Unlike Spock and the rest of us, Data has no problem with emotional control. Without his emotion chip, he has no emotions that need controlling. For him, the absence of emotions represents a huge psychological loss. And we humans can empathize with

Data. Emotions are important in our communication with everyone around us. They egg us on to appropriate action, and, as even Spock acknowledged, undergird our rational abilities.

But emotions do even more than that: they add spice to our lives. Data knows that he is missing this spice, the key ingredient that adds so much to the lives of his humanoid crewmates. And he never gives up trying to capture what he's missing. Shut out of the humanoids' rich emotional lives, Data laments:

> *I simply want to know what is funny. I want to involve myself in other people's laughter. I want to join in.*

> (TNG-2, "The Outrageous Okona")

No matter how much we may worship logic and elevate rationality (Vulcans, please take note), emotions are central to what we are, to what we do. Elation brightens success, fear darkens threat, and anger energizes intent. Chakotay, the very wise first officer of the starship *Voyager*, puts it as follows:

> *Being moved by an emotion isn't so extraneous—sometimes it's the whole point.*

> (VOY-3, "Alter Ego")

Sex, Memes, and Videotape

In the *Star Trek* universe, sex is everywhere. It oozes through every nook, cranny, and shuttlecraft, and spills over onto every holodeck. Sex may be the only thing that *Star Trek's* motley crew of alien humanoids—Vulcans, Bajorans, Cardassians, and Ferengi alike—really have in common with one another. But their shared interest doesn't mean that they all go about the business of sex in the same, one-size-fits-all style. Far from it.

Klingons are among the most passionate of *Star Trek* aliens. They engage in highly ritualized one-on-one battles, which give new meaning to the idea of "war between the sexes" (DS9-5, "Looking for par'Mach in All the Wrong Places"). Betazoids aren't as physical as Klingons, but their females go through a sexually explosive midlife period called "the phase," during which their sex drive quadruples (TNG-2, "Manhunt"). Even the normally hyper-cerebral Vulcans take the equivalent of human adolescents' Spring Break. Every seven years, an adult Vulcan male is overcome by Pon Farr, an irresistible urge to return to the home planet for sex. It's a long way to go, but apparently Vulcans think it's worth the trip (TNG-2, "Amok Time"; VOY-3, "Blood Fever").

Despite the widespread heat of sexual passion, one inhabitant of Deep Space Nine has doubts about the whole business:

ODO: I'll never understand the humanoid need to couple.

QUARK: You've never coupled?

ODO: Choose not to. Too many compromises.

(DS9-1, "A Man Alone")

So what's going on with Odo? His disinterest in sex—"coupling," in his poetic language—seems perverse. Does he know what he's missing? Odo, of course, is a changeling, a sentient, gelatinous mass that can assume any shape he desires. (Imagine a bowl of jello that can instantly turn itself into a good facsimile of your cousin Arthur, or of your family rottweiler.) While he's on duty, he assumes a humanoid form, although every sixteen hours he must revert to a puddle of orange goo to reconstitute himself. But being a shape-shifter isn't Odo's problem. In fact, his jellolike constitution allows for supercharged sensual pleasures, as we discover later when Odo and another changeling "couple" by oozing together (DS9-3, "The Search," II). To put it mildly, the ooze gets red hot.

No, Odo's problem is bashfulness and fear of rejection (DS9-5, "A Simple Investigation"). He has never been properly schooled in the complex rules of courtship. After all, sex doesn't just happen. Potential partners have to complement one another, physically as well as psychologically. And they are obligated to obey the unwritten rules of courtship. The natural history of sex has left deep, profound marks on our biological makeup and on our social institutions. However spontaneous it may seem, sex took a very long time to reach its present state of excellence.

Fortunately for Odo, he finally overcomes his fear of failure: he and a humanoid female "couple" (DS9-5, "A Simple Investigation"), with great success. Odo's four-star performance gives no hint that it's his first time. But it's a pity that Odo had to suffer on his way to sexual maturity. It would have been nice if the space station's operating manual contained a chapter entitled "Everything You Always Wanted to Know About Sex." That chapter might start by explaining the purpose of nature's ingenious invention, coupling.

Why Have Sex?

It may seem utterly irrational at times, but sex does have a purpose. No, sex is not designed just for momentary pleasure, although pleasure certainly can be a useful by-product. And its aim is not merely fruitfulness and multiplication, populating the universe with ever more of one's own kind. No, the raison d'être for sex is more important than population growth. Sex promotes biological diversity among the individual members of a species, diversity in bodily structure and in behavioral propensities. And diversity is essential for the long-term survival of a species.

To understand why this is true, consider the alternatives to sex. Frankly, if nature's only goal were to maximize the universe's population, sex would never have been invented. For sheer quantity, it's simpler and more efficient to rely on asexual reproduction, or "cloning" as it's sometimes called. In asexual reproduction, one organism (the parent) produces copies of itself (the offspring). No courtship, no coupling, just simple, Xerox-like duplication.

Asexual reproduction can work in one of several different ways. In creatures such as bacteria and amoebas, one organism splits into two, a process called *fission*. This results in a kind of two-for-one sale where, generation by generation, the products accumulate exponentially: 1, 2, 4, 8, 16, and so on—multiplication by division. The consequences of such exponential growth can be overwhelming. Take what happens on space station K-7 when Lieutenant Uhura, the *Enterprise's* communications officer, acquires a single tribble—a small, furry, purring, irresistible creature (TOS-2, "The Trouble with Tribbles"). Uhura has no idea how prolific tribbles are, or that they become pregnant merely by eating. So she brings one cuddly little creature on board the *Enterprise*. It doesn't take long for things to get out of hand. As the creatures feast on a huge supply of valuable grain, the tribble population explodes, doubling and redoubling at an alarming rate. With tribbles turning up everywhere, Commander Scott averts disaster by beaming the *Enterprise's* entire tribble population onto a nearby Klingon vessel. This kills two birds with one stone, not only solving the *Enterprise's* tribble population problem but also vexing the tribblephobic Klingons.

Tribble-like division is not the only way to reproduce without sex. Viruses, those tiny infectious agents that have plagued humankind for centuries, use an alternative mode of asexual reproduction: replication by exploitation. A virus contains the genetic instructions needed to reproduce itself (which are embodied in nucleic acid sequences within the virus's core). But it lacks the raw materials for executing those instructions. So the virus injects itself and its instructions into the cells of another organism. The cells of this host organism provide the needed raw materials for viral replication. There are two ways to look at this transaction between virus and host. From the host's standpoint, the virus's act constitutes infection, and its consequences can be harmful (as we are reminded when we get the flu). From the virus's standpoint, however, infection is essential for its survival, and the discomfort to its host is quite beside the point.[1]

To *Star Trek's* writers, self-replicating viruses are convenient villains. (These invisible bad guys are also very easy on the special effects budget.) For example, the whole crew of the *Enterprise* is infected by a virus that relaxes their inhibitions and alters their judgment (TOS-1, "The Naked Time"). While the crew is giving in to sexual abandon, the *Enterprise* is almost destroyed. A century later, Captain Jean-Luc Picard contemplates using a deadly virus to infect a captured Borg, who, upon returning to the Borg collective, would infect and kill all of its fellows (TNG-5, "I, Borg"). Picard's crew talks him out of this plan, arguing that his actions would constitute genocide. In "Chain of Command," (TNG-6, I) the Cardassians are rumored to be developing a potent virus capable of wiping out all life-forms within an ecosphere, then dissipating rapidly, leaving the territory ripe for invasion. (This "metagenic" weapon is never deployed, thanks to the sabotaging efforts of the *Enterprise*-D crew.) Aboard Deep Space Nine, crew members are infected with a virus that targets the speech areas of their brains, interfering with their verbal communication ability (DS9-1, "Babel"). A complicated outcome, but for the viral masters of self-replication, it's all in a day's work.

Self-replication without sex is very efficient, and whatever pleasures they may sacrifice, the practitioners of self-replication enjoy terrific reproductive success. So why do humans and so many other creatures bother with sex? To paraphrase Odo, what does coupling do that asexual reproduction cannot?

Asexual reproduction creates carbon copies of the original, producing offspring that are essentially identical to their single parent. Except for the occasional mutation, the individual members of a species are all alike. In contrast, sexual reproduction shuffles the genetic cards, promoting biological diversity among the members of the species. A sexually reproduced individual receives some genes from its mother and some from its father, which recombine to produce a new, unique set of genes different from that of either parent. Thus, sex ensures that offspring *won't* be identical to their parents.

But what's wrong with mimicking your parents? After all, they've survived and reproduced, so they must be doing something right. Why not stick with a proven design? Think of it this way. No species member can claim to be the only—or even the best—solution to all the challenges to the species' survival. At any one moment, each species member represents just one potential solution to those challenges. Because times and conditions change, there's no guaranteeing that any particular solution will be a lasting one. So it's best if a species can keep its biological options open, which is what sex does.

From a biological standpoint, success doesn't mean wealth or fast cars: it means successful reproduction, transferring at least some of your genetic legacy to the next generation. We often speak of success in terms of survival, but in this context, survival is important only as a prerequisite for reproduction.[2] Any characteristic—structural or behavioral—that gives you an advantage in the reproduction game will be rewarded. But the environment changes over time, and with it the rules of the game. Characteristics that are rewarded now could become burdens later. A characteristic that favors your survival may not work for your offspring.

Suppose all reproductively active female humanoids in the universe were brainwashed into believing that males with huge ears were particularly attractive, desirable sexual partners. The obvious winners would be male Ferengi, whose ear size is legendary. In this brave new universe, Ferengi males would enjoy more opportunities for reproductive encounters and would father more offspring. Moreover, the male offspring of those encounters would be favored if they too possessed large ears. Over successive generations, males with larger ears would enjoy more mating opportunities. The incidence of large ears, a genetically determined characteristic, would tend to increase in the population.

Would the advantage of large ears be maintained throughout all future generations? Probably not. If history and the laws of probability are our guides, some environmental catastrophe—say, a plague of ear mites—would eventually strike these large-eared Don Juans. At that point, the evolutionary scales would tip in favor of small-eared individuals, whose ears would be less hospitable to ear mites. Over successive generations we'd see a wholesale decrease in ear size, a trend accelerated by a growing female sexual preference for small-eared males.

The moral of this story? The environmental forces that guide evolution are not static—they change with time. In fact, many of the challenges faced by our own hominid ancestors are no longer problems for us. Our ancestors, for example, needed sharp, clawlike fingernails to shred meat and crack open nuts; fingernails probably also came in handy for self-defense against potential predators. We made clawlike fingernails obsolete by inventing knives, nutcrackers, and alarm systems. Environmental challenges, by the way, do not include just food and voracious predators. Threats to our survival include new, more potent viruses, increased radiation due to holes in the planet's protective ozone layer, shifts in water levels caused by fluctuations in global temperature, and the chronic stresses of modern urban societies. To successfully confront such emerging challenges, our distant offspring cannot look and act exactly like us. Change is adaptive and sex creates change.

Sexually reproduced individuals possess unique combinations of genes, different from those of either parent. By chance alone, some sexually reproduced individuals tend to be better equipped than others to meet life's new challenges. Those individuals with superior survival abilities are more likely to pass on their genetic legacy to the next generation. And because their superior abilities are embodied in that genetic legacy, their offspring, too, will enjoy a better chance of survival and a prosperous reproductive career.

Ironically, threats to survival are an important engine of evolutionary change. A world without threats may sound great, but there's a good chance that there'd be a steep long-term price to pay. Consider, for example, the Ocampa of the Milky Way's Delta Quadrant. When the surface of their home planet becomes so arid that it can't support life, the Ocampa survive by moving into an underground city.[3] Their underground safe haven provides for their every need. They are so well provided for, however, that they begin to lose their intellectual abilities (VOY-1, "Caretaker"). A similar fate befalls colonists on Omicron Ceti III, a planet whose atmosphere contains deadly berthold radiation. Fortunately, radiation-absorbing spores in the atmosphere protect the colonists (TOS-1, "This Side of Paradise"). But these lifesaving spores also affect the colonists' brains, making them content and passive. Threat isn't all bad—it picks the genetic lottery's winners and losers, ensuring the fitness of survivors and their offspring.

Besides promoting genetic diversity, sexual reproduction has another advantage: it weeds out harmful genetic mutations. (A *mutation* is a random "error" that alters the instructions contained within a gene.) Among our species, the mutation rate is so high that an average person will undergo about a hundred mutations during his or her lifetime. And many mutations, such as those associated with certain forms of cancer, are dangerous. It's in our best interests to weed harmful mutations out of our gene pool so that they are not passed along to our offspring. When a harmful mutation occurs in an individual that reproduces *a*sexually, it becomes a fixed part of that individual's genetic legacy. Generation after generation, every single one of that individual's cloned offspring will inherit its parent's mutations. With sexual reproduction, however, some mutations are suppressed by the recombination of genes from mother and father—the offspring may have fewer harmful mutations than either parent.[4]

Sexual reproduction promotes diversity and puts the brakes on the spread of mutations. Although we don't stop to think about it, these are the real reasons why we humans "choose" to couple. By repeated shuffling of the genetic deck, sexual reproduction produces winners and losers. The winners live long enough to play another hand of sexual poker, and the losers take their unfit genetic combinations away with them as they leave the table.

Those Ubiquitous Star Trek Humanoids

Think about the countless alien species that inhabit the *Star Trek* universe. They include some bizarre creatures, including the Horta, sentient beings made of rock that scuttle about in caves on Janus VI (TOS-1, "Devil in the Dark"); the Corvan gilvos, driftwoodlike animals living in the rainforest canopy on the planet Corvan II (TNG-5, "New Ground"); and the Crystalline Entity, a huge, snowflake-shaped life-form that sucks the energy from biological life-forms (TNG-5, "Silicon Avatar"). All very remarkable creatures. But what is truly remarkable is the number of "humanoid" species found throughout the universe—aliens who walk upright on two feet, possess arms and hands, and have mouths, noses, and eyes resembling ours. How can this be? Imitation may be the sincerest form of flattery, but nature strives for diversity, not flattery. After all, among Earth's current inhabitants, there are only a couple of primate species—ours included—that are bipedal and manually dexterous. If these qualities are so rare on present-day Earth, how did so many humanoid life-forms with the same qualities end up scattered throughout the other class-M planets of the universe?[5] Where did all these human look-alikes come from?

Our own evolutionary history provides a clue to this puzzle. Most accounts of human evolution begin some 12 million years ago in the heart of Africa, where small arboreal primates enjoyed a relatively luxurious life safe above ground in dense tropical forests.[6] The treetops provided a ready source of fruits, leaves, and nuts, and within the dense foliage of the jungle canopy, threats to their survival were few. Like the Ocampa underground haven, the forest met their every need. Then global climatic changes altered the African landscape, turning large areas of lush forests into dry grasslands. This dramatic alteration in our early ancestors' habitat forced them to abandon arboreal life for a far more difficult existence on the ground. No longer was food within arm's reach; new food sources had to be discovered, and hunting and foraging skills had to be developed. The forced move to the ground also placed our primate ancestors within easy reach of carnivorous predators.

Nature knows only one way to deal with predicaments like these. It rolls the genetic dice over and over, leaving it up to genetic recombination to produce viable solutions. Our ancestors, called *hominids*, were one of those solutions.[7] Natural selection encouraged a number of useful adaptations. One was bipedal gait. Walking on two limbs rather than all four liberates the hands for tool and weapon use. A second adaptation was keen vision, which provides advanced warning of impending threats. And a third was monogamous male-female bonding, which produces a team of committed specialists who divide up the labors of survival between hunting and child rearing. The evolutionary development of these and other useful adaptations was neither calculated nor planned. These adaptations grew out of genetic recombination guided by nothing more than the laws of chance. And they remained in the gene pool only if they bestowed on their owners some extra advantage in the game of life.

Because sexual reproduction operates by the laws of chance, it is extremely unlikely that any species is actually the best of all possible solutions to some set of environmental challenges. Humbling as the truth may be, *Homo sapiens'* design exists because it worked, not because it's the best of all possible designs. Suppose some *Star Trek*-like temporal rift rewound history 12 million years back to the period of our arboreal ancestors, erasing all the events that brought our species to its current state. If the evolutionary game were replayed from that point in time, the resulting sequence of events, and our adaptations to those events, would surely be different. You cannot replay evolution and expect to reach the same outcome.[8] To tell the truth, we are the product of numerous biological accidents. We're here today only because our ancestors won the reproductive lottery. The losers' legacy lives only in the fossil record.

Which brings us back to those humanoid species living on so many other class-M planets. Most of us have no trouble imagining that extraterrestrial life exists. Indeed, it's inconceivable that Earth is the only planet in the universe where life-forms exist. But do those other life-forms include beings that closely resemble humans? That's extremely unlikely. Alien life-forms would face very different environmental pressures from the ones that shaped the evolution of *Homo sapiens*. The gravitational forces on different planets vary. Temperatures differ, depending on atmospheric conditions and proximity to a solar source. And the variety and abundance of food sources undoubtedly differ. These planetary characteristics would shape the life-forms that evolved on each planet. Different planets, therefore, would promote the evolution of dramatically different life-forms.

And even if all class-M planetary environments *were* exactly the same, down to the last atom, the genetic dice would never yield exactly the same sequence of rolls over the several billion years needed to get from single-celled organisms to complex humanoids. The fact that so many of *Star Trek's* different humanoids look so much alike seems to violate the inviolable laws of probability.

So how did it come about that *Star Trek's* humanoids resemble one another? The answer brings us back to Odo's old nemesis: sex.

Alien Sex

Star Trek is replete with the hybrid offspring of extremely mixed marriages. B'Elanna Torres, *Voyager's* chief engineer, is living proof that Klingons can interbreed with humans. The ship's counselor aboard the *Enterprise*-D, Deanna Troi, is the product of a Betazoid/human "coupling." Tora Ziyal, Gul Dukat's illegitimate daughter, confirms that Cardassians can mate with Bajorans (DS9-2, "Indiscretion"). And we know that Vulcans can sire children with humans (Spock, TOS) as well as with Romulans (Lieutenant Saavik, *Star Trek II: The Wrath of Khan*).[9]

These successful matings among diverse *Star Trek* humanoids point to a remarkable degree of biological and behavioral compatibility. After all, several things must happen before a successful mating can occur. First of all, males and females must find each other attractive, and for this to happen, males must be able to recognize females, and vice versa. If males and females looked alike, sexual reproduction might never get off the ground. In fact, that's exactly the predicament of the J'naii, humanoids who have renounced mating as a means of reproduction (TNG-5, "The Outcast"). To discourage mating, of course, the first requirement is suppression of sexual attraction. How do the J'naii accomplish this? Simple. The J'naii are all similar in size, dress, appearance, and behavior. It's impossible to tell males from females. Sound pretty dull? It certainly does to Commander William Riker when he discusses it with a J'naii pilot named Soren. While the two of them are working together on a shuttlecraft, Soren expresses curiosity about human sexual practices. Riker begins by explaining that human males and females differ in size and bodily characteristics, including sexual organs. Soren is intrigued by this revelation:

> SOREN: Commander, tell me about your sexual organs.
>
> RIKER: Ahhh . . .
>
> SOREN: Is that an uncomfortable subject for humans?
>
> RIKER: It doesn't tend to be a topic of casual conversation.
>
> SOREN: I'm interested in your mating practices. What is involved with two sexes?
>
> RIKER: [Seemingly distracted by incoming shuttle signals] Correcting course—zero two one mark zero.
>
> SOREN: [Trying to bring the conversation back to the topic] Mating?
>
> RIKER: Right. Well, it's pretty simple. The men inseminate the women and the women carry the baby.
>
> SOREN: Our fetuses are incubated in fibrous husks, which the parents inseminate. From what we know of other species our method is less risky . . . and less painful.
>
> RIKER: And less enjoyable.
>
> SOREN: Less enjoyable?
>
> RIKER: For humans the sexual act brings a closeness and intimacy. It can be a very pleasurable experience. [brief pause] Inseminating a husk!

But other than the J'naii, females and males of *Star Trek's* various humanoid species have little trouble telling each other apart. Most of the features that differentiate the sexes in humans—

physique, hair length, voice pitch—appear in most other humanoid species, too.

But appearances will get you only so far—you have to let potential partners know you're interested. This is the step that Odo found so difficult. Males and females must send and receive compatible sexual signals, notifying each other that sex is on their minds. As we shall see, these mating rituals can be complicated, and cultural differences can get in the way. Suppose, for example, you want to lure a Klingon into the bedroom. The little signals that are common in some of Earth's cultures—flowers, dinner, a bottle of wine—just won't cut it with a Klingon. Forget the flowers; just bare your teeth, give a long, loud growl, and bite your potential mate on the cheek. Extra points if you draw blood (TNG-2, "The Emissary"; TNG-4, "Reunion"). And if your message does get through, be sure your medical insurance is paid up. The Klingon mating ritual can be dangerous. Jadzia Dax, science officer aboard Deep Space Nine, enjoys her romantic encounters with Worf (DS9-5, "Let He Who Is Without Sin"). But she pays a price, as Constable Odo wryly notes:

> DAX: [stretching her neck and grimacing] It's . . . it's just a muscle pull.
>
> SISKO: What's that? The eighth muscle pull this month?
>
> ODO: Actually, I believe Commander Dax has been treated for seven muscle pulls, two contusions, and three cracked ribs. The only person who's spent more time in the infirmary over the past month is Commander Worf.
>
> SISKO: [to Dax] Isn't there any way the two of you could . . . hmmm . . . hmmm, you know . . .
>
> DAX: Make love?
>
> SISKO: Without injuring yourselves?
>
> DAX: Interspecies romance isn't without its dangers. That's part of the fun.

Dax is clearly a consenting adult. She understands Klingon sexual practices, and she's a willing, enthusiastic participant in this mayhem. But if Dax's affections were redirected away from her Klingon lover, the sexual signals and foreplay would need to take a new form. For example, if Dax had eyes for a Bajoran, she'd have to can the growling and biting and take a more subtle, spiritual approach (DS9-2, "Shadowplay").

Once partners exchange and correctly interpret sexual signals, coupling, to use Odo's word, can take place. But coupling doesn't guarantee reproductive success. One huge hurdle remains: Successful sexual reproduction requires that the male and female participants bring common genetic blueprints to the bedroom. And here's where alien sex meets its toughest challenge. Is it really feasible that Klingons, Bajorans, Vulcans, and all the other interbreeding humanoids in *Star Trek* are genetically compatible?

For male and female humans, things usually work out rather nicely. Every cell in our bodies contains 46 chromosomes (except for sperm cells and egg cells, which have half that number). Under the microscope, these chromosomes look like long threads. Along the length of each thread lie sequences of DNA (deoxyribonucleic acid). These DNA sequences are the genes, the physical units of heredity. Each of us has approximately 100,000 genes distributed over our 46 chromosomes. These genes specify protein structure and, hence, the structural and biochemical composition of a person. To draw an analogy, each gene is a blueprint that specifies how to build a protein. Clusters of these blueprints are grouped together on a chromosome; thus each of the 46 chromosomes houses one volume of the entire set of 100,000 blueprints. Together, these blueprints—the genes—specify the materials and design for an enormously complex structure: the human body. Each gene has its own very specific role. There is one blueprint, for example, guiding the synthesis of a protein that goes into your hair; another blueprint guides the synthesis of a digestive enzyme secreted in your stomach.[10]

The successful construction of a human body requires the presence of all 100,000 blueprints. Your blueprints are a joint gift from your biological mother and father. Of the 46 chromosomes in your body's cells, 23 came from an egg cell (thanks, Mom) and 23 from a sperm cell (thanks, Dad). At conception, when a sperm fertilizes an egg, these two sets of 23 chromosomes come together to create the full complement of 46 chromosomes.

So, sexual reproduction reassembles the complete set of blueprints necessary for constructing a human being. But for the project to succeed, the male and female contributions must be compatible in number and in purpose. If even a single chromosome is missing, the offspring cannot survive—for that matter, even a few missing or defective *genes* can produce serious developmental disorders. Likewise, just one extra chromosome produces serious complications, and more than one dooms the whole project.[11]

That's one reason humans cannot successfully mate with other species on Earth—the numbers of chromosomes don't match.

But even if another species also carried 46 chromosomes, there's no guarantee that human coupling with that species would produce offspring. Actually, the chances are slim to none. For one thing, there'd have to be a very good match between each species' arrangement of genes along each of the chromosomes. The construction process wouldn't get very far if a blueprint in Volume 18A (inherited from a female of one species) dealt with the design of a protein essential to the kidney, but the blueprint at the same location in Volume 18B (inherited from a male of another species) dealt with a protein component of fingernails.[12]

Returning to the Klingons, Bajorans, and Vulcans, *Star Trek's* humanoid creatures could successfully interbreed only if the DNA sequences on each chromosome of each species were genetically compatible. Statistically, it's unimaginable that this degree of genetic compatibility could arise independently—by chance—on planets isolated from one another for much of their histories. So how *did* the same set of blueprints manage to show up in the inhabitants of so many widely scattered homeworlds?

An explanation for this unlikely compatibility comes when a congregation of Klingons, humans, Cardassians, and Romulans gathers on the planet Vilmor II. Here the group receives a message from one of the Preservers, an ancient race:

> *Our scientists seeded the primordial oceans of many*
> *worlds, where life was in its infancy. The seed codes*
> *directed your evolution toward a physical form resembling*
> *ours: This body you see before you, which is of course*
> *shaped as yours is shaped, well, you are the end result.*
>
> (TNG-6, "The Chase")

The nature of these "seed codes" is revealed when the omnipotent being Q transports Picard back in time to the moment when life on Earth is about to begin (TNG-7, "All Good Things . . . "). Q and Picard stare at a sea of soupy liquid heated by molten lava. Q explains what they're watching:

Q: Welcome home.

PICARD: Home?

Q: Don't you recognize your old stomping grounds? This is Earth—France, about three and a half billion years ago, give or take an eon or two.

PICARD: Is there a point to all of this?

Q: Come here, there's something I want to show you. See this? [The two of them are looking at a pond of liquid ooze.] This is you. I'm serious. Right here life is about to form on this planet for the very first time. A group of amino acids are about to combine to form the first proteins, the building blocks . . . of what you call life. Strange, isn't it. Everything you know—your entire civilization—it all begins right here in this little pond of goo. Oh look, there they go, the amino acids are moving closer and closer and closer.

Q seems to think that humanoids all over the universe look pretty much alike because many worlds were seeded independently with the same humanoid DNA. It is because the inhabitants of those worlds come from common genetic stock that they have such a strong family resemblance.[13]

But even though Q is quasi-omniscient, his theory seems to be—well—mistaken. Humans could not have sprung directly from that pool of goo. Earth's primordial ooze might have supported the evolution of simple unicellular life-forms that reproduced asexually. But, no matter what Q says, it was not the spawning ground for humans, or humanoids, or anything close. Granted, the Preservers might have seeded our oceans with the ingredients for life, but those ingredients were not enough to produce humanoids. That took billions of years and untold chance occurrences. Once the first life-forms developed, evolution would work its magic all right, but very slowly, and thanks to the fickle laws of probability, not so surely. The enormous time that humanoid evolution would require, together with interplanetary variations in environmental conditions, would guarantee that evolution on different planets—even starting from the same seed DNA—would diverge far more than *Star Trek* suggests. So the account offered by the Preserver and by Q can't be entirely correct.

Perhaps with a little revision, we can rescue Q's explanation of the universewide similarities among humanoid species. The seed material the Preservers distributed throughout the universe could have been harvested, millions of years ago, from our hominid ancestors. In Africa at that time, our branch on the hominid tree was just starting to bud, and the DNA sequences of those hominids would have been very similar to the DNA sequences of present-day *Homo sapiens*. Perhaps the Preservers took those DNA sequences—highjacked blueprints for building hominids—and carried them to other class-M planets, where they were used to clone humanoids.

But even this revised theory must overcome some pretty big hurdles. Humanoid cloning could not begin to succeed unless that pirated DNA got very special attention. Just dumping it into a pool, like the one Q shows Picard, would be a total waste of good genetic material. The Preservers would need suitable host cells in which the hominid DNA's coded instructions for replication could be implemented. But not just any host cell would do—a viable host would have to possess a nucleus to house the DNA, as well as the requisite raw machinery for protein synthesis.

Even if viable host cells could be found on other planets, the self-replicating DNA molecules wouldn't immediately produce an adult humanoid. The multiplying cells would have to undergo a lengthy, complex period of embryonic development. That development, which is the expression of the DNA's instructions, would require an environment that mimics the fluids of the human female amniotic sac. Such an environment would be hard to find, even on the classiest class-M planet.

And even if every single one of these stringent requirements could be met, the developing humanoid would still require careful nurturing after it was born. Where did the Preservers find the nannies who fed and cared for these helpless infants? Keep in mind, too, that these humanoid clones, regardless of the quality of child care, could survive only on planets whose environmental conditions resembled those on Earth—the conditions for which our DNA sequences are adapted.

This sequence of events, implausible though it seems, could explain the presence of humanlike life-forms on other planets. But this raises another question: Why don't all of *Star Trek's* humanoid aliens look *more* alike than they do? The answer to this one is easy. While our hominid ancestors were continuing to evolve right here on Earth, our genetic kin on other planets were adapting to whatever environmental challenges their planets threw at them. A few million years is plenty of time for physical and behavioral differences to have evolved among these hominid descendants. (Consider, for example, the differences in appearance between Eskimos and Africans.) No wonder Klingons look different from humans.

Meeting of the Memes

In the *Star Trek* universe, some humanoid species developed superior intelligence much more quickly than others. Six hundred thousand years ago, when our human ancestors were still figuring out how to cook with fire, the Vulcans and Romulans had already developed the technology for interstellar space travel (TOS-2, "Return to Tomorrow").

Assuming that some version of the Preservers' story is true, humans, Vulcans, and Romulans all got started at roughly the same time. How, then, did the others leapfrog ahead of us intellectually? Did evolution accelerate the growth of their brains compared with ours? Not necessarily. Brain size is not the be-all and end-all of intelligence. Once a species develops the ability to manipulate symbols and think conceptually, brain size per se is no longer the limiting factor. After all, our own species managed to land men on the moon, even though our rocket scientists had to make do with the same brains that a century earlier had struggled to create the sewing machine. Actually, space travel may sound more complicated than it really is. Yes, it's fraught with technical and logistical problems. But with a handful of key breakthrough discoveries—jet propulsion, transistors, and, of course, powdered orange juice—any humanoid species with a brain of reasonable size and sufficient resources could be on its way to the stars.

There's an important lesson here: At some point in humanoid history, biological evolution takes a back seat to cultural evolution. The evolution of physical characteristics, including brain size, is very, very slow; major modifications require countless generations. The evolution of ideas, however, isn't regulated by a developmental clock that ticks on a generational scale. Rather, ideas can develop, spread, and hybridize in the blink of an evolutionary eye. Biologist Richard Dawkins[14] uses the word *meme* (pronounced "meem") to signify an idea that is shared by members of a culture. Memes are concepts, bits of information that are replicated in the minds of different individuals. Examples of our culture's memes include jazz, flossing twice a day, mass transit, and monotheism.[15]

In a way, memes are like genes: genes are the units of physical evolution, and memes are the units of cultural evolution. Memes and genes differ, however, in the way in which each unit replicates. Genes replicate by passage from body to body through the gametes (sperm and egg); memes replicate by passage from brain to brain through social communication, by spoken and written word and by "human-see-human-do" imitation. Within an electronically interconnected world like ours, memes spread like wildfire.

Memes have awesome power. They ignite wars, as when the Cardassians get the notion that space station Deep Space Nine should again be under their control (DS9-5, "By Purgatory's Light"). They cure diseases, as when the Ornarans realize that they will not die just because they run out of felicium, a highly addictive drug they mistook for life-saving medicine (TNG-1, "Symbiosis"). Our species has reached the stage at which most major advances—and most catastrophes—will be spawned not by evolution, but by new memes and new meme combinations. Evidently the Vulcans and their rebel cousins the Romulans discovered the power of a few good memes earlier than we did.

Quantity versus Quality: The Sex Equation

Whatever power memes exert in electronic, interconnected societies, it was sex that brought those societies to the point where they are now. The equation for sexual reproduction is simple: A gamete (a single cell containing genetic material) produced by one parent fuses with a gamete from the other parent, forming an offspring that contains genes from both parents. In some simple organisms, the two parents' gametes are indistinguishable in size. But in higher organisms, including *Homo sapiens*, the gametes have evolved into two types, and the differences between these two types have very practical consequences. Had Odo only understood these consequences, he might have succeeded more quickly in the mating game.

One type of gamete consists of large, nutrient-rich cells, called eggs, that are produced in very small numbers; the other type consists of very small gametes, called sperm, that are produced in astronomical numbers. In the human species, the female releases a single large ripe gamete every twenty-eight days or so throughout her reproductive years. If you take into account gestation time and postnatal nursing, no human female could produce more than a few dozen offspring during her lifetime. Her male counterpart knows no such limit. With little effort, he manufactures approximately five million gametes per day, every day, year in and year out, over decades of his life. And this army of sperm is ready for instant deployment whenever opportunity arises. One man alone could double Earth's population in just eighteen months if every one of his sperm produced an offspring. The prospect—and the college tuition bills—are frightening.

Understanding this, you can imagine why females and males might approach sex differently. Because her reproductive opportunities are so limited, a biologically prudent female will shop carefully before selecting a male with whom to share the experience. She will seek a mate whose genetic contribution will strengthen her offspring's chances of survival. Males, in contrast, can be more casual about their sperm. For every million sperm wasted, a nearly inexhaustible supply of replacements waits in the wings. Because sperm cost so little to make, males can maximize their reproductive success by mating every chance they get.

Phrases like "shop carefully" and "mating every chance they get" don't mean that females and males consciously realize which reproductive strategy best serves their particular interests. These strategies, which evolved millions of years ago, entail neither thought nor reason, but are as automatic as the way we tend to smile when someone else smiles at us.

A female's egg is made even more precious by the length of her pregnancy and by the time she will invest in nurturing her offspring. In a number of species, including our own, the female keeps the egg inside her body for an extended period of time after fertilization. During this lengthy pregnancy, the mother is literally eating for her fetus. And once it is born, the mother's life continues to revolve around nourishing and raising her offspring. She invests a huge amount of time and energy in her child.

From a biological perspective (certainly not the only perspective that counts), a female's investment can't pay dividends until her child reaches reproductive maturity. At that point in its development, it is finally in a position to pass on its mother's genetic legacy. Raising a child to the point of sexual maturity represents one of life's great burdens, one borne disproportionately by the mother. As one biologist put it, "From the female's point of view the male is little more than a parasite who takes advantage of her dedication to reproduction."[16]

For human females who feel exploited by males in the reproduction department, *Star Trek* offers a vision that promises a solution. On the planet Taresia in the Delta Quadrant, the tables are turned (VOY-3, "Favorite Son"). Here, females outnumber males nine to one, and there's a good reason why: males are dying for sex—literally. For starters, males are lured to the planet with a DNA-altering virus, which makes them behave like salmon returning to their home waters to mate. Once he gets to Taresia, the male is pampered and fawned over by a squad of young, nubile females, and is told that all he has to do is choose which three squad members will marry him and attend to his every sexual wish. Once the brides are selected, a formal ceremony seals the deal. Then it's on to the honeymoon night, which is a real killer.

Anticipating a night of bliss, the male eagerly puts himself in the hands of his three brides. They repay his trust in an unusual way: they harvest his internal organs, leaving the bridegroom a mere shell of his former self. The self-made widows extract DNA from their late husband's tissues and use it to fertilize their eggs. To make this system work, the Taresian ladies have to harvest a great many cells—and also a great many husbands.

Fortunately for males, Taresian sexual practices seem to be the exception—not the rule—in the universe. Most of *Star Trek's* humanoids engage in a more balanced, far less lethal mating game.

The Mating Game

> *Animals—including humans—spend an inordinate amount of time getting ready to have sex. Something that could be achieved by mutual agreement in a minute or two is commonly drawn out into hours, days, even weeks of assiduous pursuit, comical misadventure, and brain-numbing stress. In a word: courtship.*
>
> S. LeVay (*The Sexual Brain*, 57)

> Q: These mating rituals you humans engage in are quite disgusting.
>
> (DS9-1, "Q-Less")

For sex to happen, males and females must follow certain rules that guide their steps toward a sexual encounter. Q may think that human mating rituals are disgusting, but those rituals are necessary for successful mating. And it was those rules that stumped Odo for so many years. Perhaps Odo should have spent more time in Quark's, for a friendly bar turns out to be an excellent place to see the rules of courtship acted out.[17] Just ask Lal, Data's newly created android daughter.

Lal is sitting with Guinan in Ten-Forward, the *Enterprise*-D's bar and lounge. Under Guinan's tutelage, Lal is learning the rules of heterosexual intimacy. With great seriousness, Lal studies a nearby pair of humans, each deeply interested in the other (TNG-3, "The Offspring"):

LAL: What are they doing?

GUINAN: It's called "flirting."

LAL: They seem to be communicating telepathically.

GUINAN: They're both thinking the same thing, if that's what you mean.

LAL: Guinan, is the joining of hands a symbolic action for humans?

GUINAN: It shows affection; humans like to touch each other. They start with the hands and go from there.

LAL: [In an agitated voice] He's biting that female!

GUINAN: No, he's not biting her. They're pressing lips—it's called kissing.

LAL: [Watching the couple leave Ten-Forward] Why are they leaving?

GUINAN: [Rolling her eyes] Some things your father is just going to have to explain to you when he thinks you're ready.

Like Lal, some social scientists spend time in cocktail lounges in order to observe male-female interactions and to deduce courtship's rules.[18] They look for the same nonverbal cues that Lal picked up on, including eye contact, proximity, and gestures involving physical contact. The flow chart on the facing page summarizes the highly predictable sequence of behaviors called courtship.

Individuals—males and females alike—begin courtship by drawing attention to themselves. In this "look at me" stage, capturing attention is key. Males often engage in exaggerated actions, such as waving their hands during conversation, talking loudly, and fidgeting a lot. Females spend a lot of time smiling, shifting their gaze around the room, and fiddling with their hair. Males and females move about with characteristically different walks, which accentuate each one's "best parts."

Now we come to a key stage in the process: mutual eye contact. At first, shared gazes last only a few seconds, and they're accompanied by coy turns of the head and subtle movements of the mouth and lips. Looking someone else in the eyes sends a very strong signal that demands a response. It's the opening move in a game that has been played countless times throughout animal history. Mammalian females most readily initiate eye contact when they're in heat (during ovulation, when estrogen levels are high). The male's response also depends on his hormonal

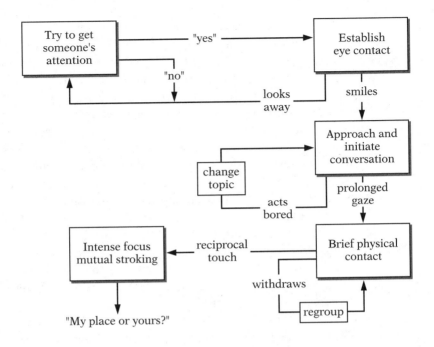

state—eye contact is more readily reciprocated if he has not recently had a sexual encounter.

Even the sexual novice Odo scores high marks on this eye contact test—a beautiful female stranger in Quark's bar seductively compliments him on his "bedroom eyes" (DS9-5, "A Simple Investigation"). And he wasn't even trying! But not all Starfleet personnel are so blessed. In fact, courtship's eye-contact demands put Geordi LaForge, the engineer aboard *Enterprise*-D, at a distinct disadvantage. Blind from birth, LaForge wears a device called a VISOR that provides him with visual stimulation (we shall examine the VISOR in greater detail in Chapter 5). The VISOR covers his eyes, ruling out any possibility of direct eye-to-eye contact. His inability to gaze longingly into the eyes of a potential mate probably contributes to his difficulties in courtship and to his bashfulness in romantic situations (TNG-3, "Transfigurations"). Over drinks in the bar on the *Enterprise*-D, LaForge confesses to Worf that he is attracted to Christy Henshaw, a female ensign on board the starship. Alas, LaForge doesn't know how to get things started. Assuming the role of LaForge's coach, Lieutenant Worf stresses that it is imperative to let a woman see "the fire in your eyes." Good advice for most people, but Worf seems oblivious to LaForge's disability.

Getting back to the courtship ritual, once mutual eye contact is made, what determines where things go next? Depends on where you want them to go. To nip courtship's progress in the bud, a disinterested party simply breaks eye contact, often with a slight downturn of the mouth that insinuates mild disgust. These actions constitute a turnoff signal, bringing the brief exchange to a grinding halt. But what if you want to signal continued interest? All that is required to keep things going is a smile. This acknowledgement of mutual interest leads both parties to move closer, within speaking range. The opening conversation typically revolves around benign subjects, such as occupations. During this conversation, the two individuals gradually turn to face one another. At first, it's head only, then gradually the rest of the body joins in.

This brings the interaction to a second key juncture: physical contact. Contrary to stereotype, it's often the female who initiates the first touch. A light stroke on the male's arm or shoulder is the female's way of reaching out. Males, in comparison, are less subtle. When a man initiates first contact, his action tends to be bolder and more prolonged: he rests his hand on her arm or her hand. At this stage, the person touched either condones the action by facial expression or reciprocal touching, or terminates the touch by moving away. It's at this stage of courtship that the Ferengi Quark routinely strikes out. He is infamous for placing his hand on the thigh or buttocks of any female he's attracted to, a maneuver that invariably leads to rejection and threats of retaliatory dismemberment (DS9-3, "The House of Quark"). In defense of Quark's persistence, we must understand that his behavior reflects his Ferengi heritage—males in this species dominate females. In addition, Quark's aggressiveness in courtship is part and parcel of his unabashed courtship philosophy: "What's love without danger?" (DS9-5, "The Ship").

If touching is encouraged and reciprocated, the committed parties graduate to greater intimacy. They stare into each other's eyes for prolonged periods, and they gaze at each other's mouths and hair. At this stage, the two begin to move in synchrony, lifting glasses together, stroking hands in rhythm, ultimately reaching what romance researchers call "full-body synchronization."[19] During this period of intimate synergy, the partners are oblivious to their surroundings—they only have eyes for each other. Their synergy often culminates in public kissing (the action Lal mistook for biting). Further escalations in physical intimacy are deferred until privacy can be achieved.

This romantic crescendo is not unique to bars. Anthropologists describe comparable interactions between potential mates in other social settings and in diverse cultures around the world.[20] The similarities extend to nonhuman animals and, according to *Star Trek*, to various humanoids. When Quark is looking for tips about how to woo a Klingon female, he's advised to give her a gift of food and then get her talking about her family (DS9-5, "Looking for par'Mach in All the Wrong Places"). Quark may be a Ferengi, and his intended may be a Klingon, but the advice he got could have come from right here on Earth.

What Your Brain Does During Sex

In humans, the brain is the primary sex organ. Within that organ, one part in particular—the *hypothalamus*—is intimately involved in the control of sexual behavior. Located deep within the interior of the brain (see the illustration below), this grape-sized structure is made up of several dozen or so nuclei (a nucleus is a cluster of neuron cell bodies). Although the hypothalamus constitutes less than 1 percent of the brain's total volume, its various nuclei play

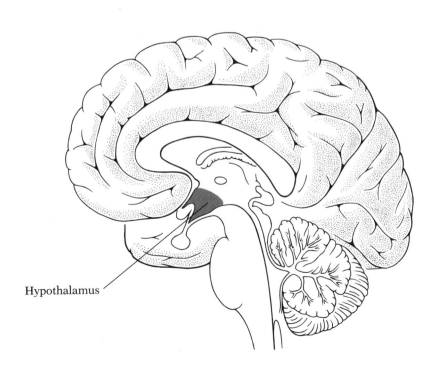

Hypothalamus

crucial roles in the regulation of bodily functions such as eating, drinking, cardiovascular function, aggression, stress reactions—and, of course, sex. The hypothalamus is a unique part of the brain in one important respect: its neurons, besides sending and receiving electrical messages by conventional means, are also exquisitely sensitive to direct chemical stimulation by hormones circulating in the bloodstream. (Hormones are chemicals secreted by the various glands of the endocrine system.) In addition to responding to hormones, some neurons in the hypothalamus secrete hormones of their own. This glandlike ability to respond to and secrete hormones enables the hypothalamus to fuel and direct sexual arousal.

Young Wesley Crusher experiences the driving force of sexual hormones firsthand when he falls head over heels for Salia, a cute young female who visits the *Enterprise*-D (TNG-2, "The Dauphine"). Geordi LaForge comforts Wesley by pointing out that infatuation is commonplace at Wesley's age because the glands are bursting with hormones. (Unfortunately for Wesley, his infatuation leads to heart-break when he discovers that Salia is a shape-shifter whose normal form is so hideous that it would turn off the glands of even the most randy teenager.)

When it comes to actual coupling, Starfleet personnel are dis-creet. True, we've witnessed their building passions during courtship, and we've seen their post-coupling afterglow. But we've never actually seen Starfleet personnel engaged in sex. What do you suppose goes on in their brains during the act? For that mat-ter, what goes on in our brains? That question remains prudently unanswered. But if our primate ancestors are any indication, sex is accompanied by a hypothalamic thunderstorm. Neuroscientists have recorded neural activity in this region of the brains of mon-keys just before and during copulation. One particular cluster of cells in the hypothalamus (termed the medial preoptic nucleus) discharges furiously when a male monkey sees a receptive female. During copulation, these cells fire less frequently, and following ejaculation, they are completely inactive.

These hypothalamic cells are not simply responding indiscrim-inately to any old exciting prospect. When a male monkey sees an appealing but nonsexual object such as a banana, cells in this brain region are nonchalantly inactive. The medial preoptic nucle-us is sexually charged because its cells are loaded with receptors that readily latch onto testosterone, one of the male sex hormones. Increases in the level of testosterone in the blood sensitize these neurons, making them easier to activate and, consequently, boosting the male's sex drive.

While we don't advise trying this at home, it is possible to jump-start a male monkey's sexual sequence with a little jolt of electricity to the right place in his brain. In a quiet, unaroused monkey, electrical stimulation of the medial preoptic nucleus instantly rivets his attention on sex. The male immediately searches for a receptive female and attempts to copulate with her. This hypothalamic "on" switch gives new meaning to the phrase "a turn-on." And as you might expect, surgical removal of this area dulls the monkey's interest in sex. Surgery has the same dulling effect in human males. The medial preoptic nucleus is sometimes removed as a treatment of last resort for repeated sex offenders.

The picture is clear: The medial preoptic nucleus is a major brain site for control of sexual interest and sexual behavior in males. In females the brain's comparable "hot spot" is a few millimeters farther back in the hypothalamus, in the ventromedial nucleus. Same story, different nucleus: In female monkeys, electrical stimulation of this brain region instigates sexual behavior, its destruction thwarts sexual behavior, and its neural activity goes up preceding and during copulation.

A warning is in order: Neither of these regions can legitimately claim to be *the* "sex center." In both males and females, cells in the hypothalamus send their signals to many different parts of the brain. And these cells receive neural inputs from a host of other brain areas, including, for example, sensory areas registering information about smell.[21] Humans understand odor's power to stimulate sexual interest, and we spend millions each year on perfume, cologne, and other smell-enhancing aphrodisiacs. The android Data is puzzled by the ability of perfume to produce sexual arousal in his human crewmates, and he wonders what smell has to do with sex (TNG-1, "Angel One"). The answer is right under his nose. Neural signals from the brain's olfactory cortex (the brain area responsible for odor perception) act like a television set's volume control: they modulate activity in the hypothalamus, raising or lowering the intensity of sexual responses. A splash of Chanel No. 5 cranks up the sexual volume, while a stopped-up nose pushes the sexual mute button. The changeling Odo, by the way, has no sense of smell whatsoever (DS9-1, "If Wishes Were Horses"), a condition that is known in humans as anosmia. It could be that Odo's shyness about sex is caused in part by his insensitivity to odors.

In most mammals, females are sexually receptive only around the time of ovulation, and in some species ovulation occurs only once or twice a year. In the *Star Trek* universe, the record for ovulation infrequency must belong to the Ocampa. As we learn first-hand from the experiences of Kes, the medical assistant aboard *Voyager*, a female Ocampa enjoys one—and only one—period of fertility during her entire lifetime (VOY-2, "Elogium"). This single period of sexual "heat" is red hot, but it lasts only about two days. With mating time so short, the female had better be prepared to act quickly if she's going to be a mother. This very limited window of opportunity represents an effective form of birth control, a necessity given the extremely meager resources available on the Ocampa homeworld (VOY-1, "Caretaker"). Things aren't nearly so urgent for Vulcans, who enter a brief period of sexual heat every 7 years (VOY-3, "Blood Fever"). But by human standards, that's still pretty infrequent.

By all accounts, human females are among the universe's most sexual beings. Although women ovulate only once every month, they can be receptive to sex anytime throughout the menstrual cycle and, for that matter, even after menopause, when the ovaries are no longer producing sex hormones. Among human males, sex drive is related to levels of testosterone, which is produced by the testes. Thus castration often blunts and eventually eliminates a male's sex drive, while testosterone therapy can revive a disinterested male. As we have seen, testosterone's stimulating effect stems, at least in part, from its effect on cell sensitivity in the medial preoptic nucleus of the hypothalamus.

If successful sexual behavior depends on hormones and specialized brain areas, is it conceivable that a sentient machine could perform sexually? Contrary to common sense, the answer appears to be yes. In response to Tasha Yar's query about whether he is "fully functional," Data responds, "In every way, of course. I am programmed in multiple techniques, a broad variety of pleasuring." This answer elicits an erotic purr from Tasha and the reply, "Ahhhh, you jewel—that's exactly what I hoped" (TNG-1, "The Naked Now"). And in the movie *Star Trek VIII: First Contact*, Data reconfirms his sexual prowess when questioned by the Borg queen. Were these merely idle boasts? Toward the end of the episode "The Naked Now," Tasha Yar leads us to believe that her tryst with Data was never consummated—she tells Data, "It never happened." But was Tasha lying? After her retirement from the *Star Trek: The Next Generation* cast, actress Denise Crosby, who played Tasha Yar, was asked what Data was really like in bed. Her wry response was, "Like a food blender, able to operate at three speeds."

Jealousy and Divorce

Sometimes the carefully programmed sequence of courtship and mating goes awry. Take, for example, the roller coaster romance between Neelix and Kes on board *Voyager*. Their growing romance almost culminates in sex. At the last moment, however, Kes calls off the coupling, realizing that she's just not ready for the responsibilities of parenthood (VOY-1, "Elogium"). Not long after that, Kes and Tom Paris (the ship's testosterone champ) share an exciting moment on the holodeck that, for Paris at least, carries sexual overtones. Neelix discovers what's going on and becomes inflamed with jealousy. Unable to repress his emotions, Neelix insults Paris and, males being males, a fist fight ensues (VOY-1, "Parturition").

Jealousy has been featured in countless songs, movies, operas, novels, and even a murder trial or two. But what is this green-eyed monster that stirs such passions and hatreds? Why are humanoids jealous of one another?

One particularly interesting view of jealousy's biological origins comes from evolutionary psychology. This school of thought treats the mind as a collection of competencies that natural selection shaped in our ancestors. These competencies reflect solutions to the environmental and social challenges that our ancestors faced in the past. Jealousy is a solution that worked for them, and because we've inherited their bodies and their brain circuits, we're stuck with it. This viewpoint can't remove the sting of jealousy, but it can help us understand our reactions to infidelity, particularly the differences in male and female reactions to infidelity.[22]

Because a pregnant female carries her fetus internally, she never doubts whether the child is her own. But, without a DNA test, a male can never be certain that he's the father of a particular child. And apparently males have every right to be uncertain. Genetic studies in humans reveal that about 10 percent of children are *not* the offspring of their putative fathers.[23] This uncertainty about paternity raises a disturbing possibility: sexual infidelity by his mate might con a man into raising and supporting some other male's child. In evolution's blunt terms, a male always runs the risk of expending his resources to promote the survival of someone else's genes. Thus, jealousy in males is easily triggered by signs of sexual infidelity.

Things are different for a female. For her, there's not a moment's doubt about her offspring's genetic heritage. And her mate's sexual infidelity won't change this a bit. (This is not to say that male sexual infidelity has no emotional impact on a woman—obviously it does.) So what makes a female most strongly jealous? Putting the question in evolutionary terms, what would most jeopardize a female's genetic legacy? From a mother's perspective, raising a healthy child to the age of puberty is a huge challenge. And to meet it, she relies heavily on her mate for help. Consequently, a female should be more threatened than a male by the potential dissolution of their relationship. Therefore, evolutionary psychology predicts that a woman will react most strongly to the threat of emotional infidelity by her mate.

These predictions from evolutionary theory seem to be borne out. In surveys, males and females have been asked which they would find more distressing, a mate who fell in love with another person ("emotional" infidelity) or a mate who had sexual relations with another person ("sexual" infidelity). In a variety of countries, including the United States, the results are consistent: Males tend to say sexual infidelity, but females tend to say emotional infidelity.[24]

Of course, these sex differences might have nothing to do with evolutionary biology. Instead, males may be more worried than females that sexual infidelity will lead their spouses to abandon them. This cognitive explanation focuses on the *perceived* consequences of sexual infidelity.[25] Does society program us to think this way, or do our different ways of thinking about infidelity occur naturally because of genetic programming?

One clue comes from the well established double standard for male and female sexual behavior. In nearly all societies, women are held to a stricter sexual standard than men. Men typically have more sexual partners over their lifetimes than do women; females who engage in casual sex are frowned upon more than men who do so; males typically have their first sexual encounter at an earlier age than females; and men are more likely to engage in extramarital sex than are women. These differences remind us of that distinction between quality and quantity. Prudent females carefully screen potential mates for fitness, perhaps as Jadzia Dax screened Worf, but males are more like Captain Kirk, inclined to mate whenever the opportunity arises.

Evolutionary psychology does not assume that people are conscious of the biological forces underlying the sexual double standard. When the average person is asked why a double standard exists, the answer may be carefully reasoned, plausible, and convincing. But odds are it's also wrong. Biology's most ingenious mandates are executed outside of conscious awareness, which means that people are actually poor judges of why they really behave as they do.[26]

Not everyone believes that the sexual double standard is grounded in evolution. Another school of thought, identified with one feminist perspective, characterizes the sexual double standard as a manipulative, self-serving moral code dictated by males. It's a strategy on the part of males to limit female reproductive options. According to this view, a woman is taught to expect, and therefore to tolerate, infidelity by her mate, but to avoid infidelity herself because it could cost her the resources he provides.

Regardless of where it comes from, jealousy can be fatal to a relationship. If that relationship involves marriage, then we're talking divorce. In fact, over a wide range of societies, the two most oft-cited reasons for divorce are sexual infidelity and sterility or barrenness.[27] Can it be coincidental that both reasons are sexually based? Consider this: The incidence of divorce is highest among those in their twenties, and plummets as people get older. Remarriage ("the triumph of hope over experience," according to Samuel Johnson) is much more prevalent among those in their late twenties to mid-thirties than among older couples. Thus, it is during the height of their reproductive years that males and females are most likely to dissolve unproductive relationships and form new ones. Lwaxana Troi's divorce from her second husband, the Tavnian diplomat Jeyal, was triggered by Lwaxana's realization that she would not be allowed to participate in the rearing of their child. Perhaps not so coincidentally, this divorce occurred while she was still within her child-bearing years (DS9-2, "The Muse").

Don't get the wrong picture. Not every relationship is inflamed by jealousy, and not every marriage is doomed to end in divorce. Many human couples actually manage to get along with each other, despite evolution's seemingly incompatible agendas for males and females. This near-miracle could come out of the common sense that's embodied in many memes, those various bits of cultural wisdom about how males and females *can* get along. Or it might come from what each of us learns firsthand about ourselves and about members of the opposite sex.

Although the learning process can be painful and fraught with error, it can also be pleasurable, as Natira, High Priestess of the asteroid-spaceship *Yonada,* put it: "Is not that the nature of men and women—that the pleasure is in the learning of each other?" (TOS-3, "For the World Is Hollow and I Have Touched the Sky"). Whatever the explanation, it does seem that many *Homo sapiens* are working themselves freer from nature's biologically scripted sex roles.[28]

*K*lingons Will Be Klingons

Outside, it's a depressing, gray afternoon. But Dr. Walter Hess is oblivious to the weather. He's not even sure whether it's day or night. He is far too busy finishing a long surgery, and much too excited about what he expects to happen next. Hess has spent hours threading a thin wire into just the right spot deep in the patient's brain. Unaware of Hess's intentions, the patient sits calmly on a table. Then Hess throws a switch, and electrons race down the wire into the patient's brain, jolting thousands of neighboring neurons into action. The transformation is instantaneous: hair standing on end, pupils open wide, teeth bared, the patient hisses long and loud at an unseen enemy. Hess is delighted. Somehow, out of an ordinary stream of electrons, he has created rage and fury. Then Hess snaps off the electricity, and just as quickly as the rage came on, it disappears.

That was 1928, in Zurich, on Earth, and Hess's patient was a member of the species *Felis domesticus*—an ordinary domestic cat. Now jump ahead four centuries, from Hess's laboratory to a dungeon in the Delta Quadrant of the Milky Way galaxy. Jump also from enraged cats to even more enraged humanoids.

No prison is fun, but for nastiness and brutality, this one's off the charts. The inmates battle one another at the drop of a hat. A few scraps of food are a perfectly good reason to slit their temporary owner's throat. "He won't be needing them anyway," a murderous inmate boasts as he snatches the scraps from the hand of his dying victim. The brutalized inmates of the prison include Harry Kim and Tom Paris, officers from the United Federation of Planets starship *Voyager*. Kim and Paris have been convicted by the Aquitarian government of a terrorist bombing that killed forty-seven security officers.

They may be far apart in time and space, but the Aquitarian prison is a lot like Hess's laboratory. Just like Hess's cats, the Aquitarians' humanoid prisoners go into violent rages for little or no reason. And like Hess's cats, the prisoners' violence subsides as quickly as it comes on. There's a good reason for these similarities. In each inmate's brain, the Aquitarians have implanted an electronic device—affectionately nicknamed "the clamp"—that stimulates the same part of the brain that Hess stimulated in his cats.

After Tom and Harry are rescued from the prison, *Voyager's* Emergency Medical Hologram removes the clamps from their brains and explains:

> *The implants are designed to stimulate the production of acetylcholine in the hypothalamus. . . . Acetylcholine is a brain chemical common to the neural structures of most humanoids. Essentially, it helps stimulate one's aggressive tendencies.*

> (VOY-3, "The Chute")

The holodoctor's diagnosis squares perfectly with what Walter Hess deduced four centuries earlier:[1] The hypothalamus, a peanut-sized cluster of nuclei nestled just below the center of the brain, contains a fuse for aggressive behavior. When certain cells in the hypothalamus get riled up with neural excitement, there's a good chance that the brain's owner will explode with rage. But like conventional fuses, the hypothalamus has no power of its own. It can promote aggressive behavior, but only because it has strong ties with other brain regions where aggression is shaped and directed. These regions form a network that stretches from the top of the brain, the cerebral cortex, down to its basement, the brain stem. And on its way down, the network connects with the limbic system, including the hypothalamus. Working together, the components of this network orchestrate the intensity of aggression—and its focus.

The neural network underlying aggressive behavior is a tightly integrated system that makes it possible for our species to do its worst. Over millions of years of evolutionary history, this brain system has had its finger in countless acts of violence, murder, torture, and war. Hundreds of millions of lives, humanoid and nonhumanoid alike, have been sacrificed on aggression's altar. But why would nature create this neural time bomb? And why would nature give it a hair-trigger fuse?

Aggression's Many Faces

The word *aggression* comes from a Latin word meaning "attack." In *Star Trek*, humanoids have used just about every means imaginable to attack one another: phasers, photon torpedoes, bat'leth,[2] fists, knives, and even name-calling. (Remember that old schoolyard chant about sticks and stones?) Here's a definitely unscientific sample of aggressive acts:

• Science Officer Spock deftly uses his fingers to apply pressure to nerves at the base of Captain James Kirk's neck, knocking Kirk unconscious. Viewers of this scene are privileged to be present at the birth of the soon-to-be-famous Vulcan nerve pinch (TOS-1, "The Enemy Within").

• We're in a dimly lit bar on Starbase Earhart. A trio of huge, nasty, ill-tempered Nausicaans insults Ensign Jean-Luc Picard, who is fresh from his graduation from Starfleet Academy. When he tries to defend his honor, Picard is stabbed through the heart. His life hangs by a thread until he gets an artificial heart (TNG-2, "Samaritan Snare"; TNG-6, "Tapestry").

• Kira Nerys, second-in-command on board Deep Space Nine, shouts angry threats at the Ferengi Quark when he makes an unwelcome sexual advance by putting his hand on her hip. Kira's threat of retaliation quickly extinguishes Quark's infelicitous behavior (DS9-1, "Emissary").

• During a passionate romantic exchange, Worf bites his Klingon mate K'Ehleyr on the hand. It's not just a love nip—his passion leaves her hand bleeding (TNG-2, "The Emissary").

These four incidents from *Star Trek* differ in their details, but they also seem to have something in common. We can use them to sort out what aggression is and what it is not.

Let's start with the bar fight between Ensign Picard and the three Nausicaans. The rowdy Nausicaans are taunting Picard and two of his Starfleet classmates, calling them cowards for refusing to play a game of dom-jot.[3] Picard is angered by their taunts and takes a swing at one of the Nausicaans. Predictably, a brawl breaks out. Picard holds his own against his much bigger adversaries until one of them pulls a knife and stabs him in the back. The blade of the huge, Nausicaan-sized knife punctures Picard's heart.

Any witness to this scene would agree that the actions of everyone involved—humans and Nausicaans alike—constitute aggression. One humanoid purposely inflicts pain, injury, or distress upon another. When Picard attacks the Nausicaans, that's aggression. When one of the Nausicaans stabs Picard, that's aggression, too. On both sides of the conflict we have the essence of aggression: behavior meant to harm or injure another living being.[4] Fists or knives, it doesn't matter—intention to injure comes across loud and clear. It's also as plain as a Nausicaan knife that the intention succeeds.

Next, what about the sexual harassment incident, in which Quark gropes Major Kira? She doesn't hit him. She doesn't even make a fist. Instead, she warns Quark to knock it off. Kira doesn't have to punch him, stab him, or hit him with a two-by-four, but she certainly gets her message across: "Keep up this behavior and you'll be one sorry Ferengi." The threatening tone of her voice and words comes across loud and clear. This too qualifies as aggression: the threat to inflict bodily harm.

But what about Spock's nerve pinch? After all, his action stopped Kirk cold in his tracks, rendering the captain totally unconscious. And the pinch was intended to do precisely that, whether Kirk liked it or not. But Spock had no intention of harming his friend—the nerve pinch was Spock's way of subduing Kirk so the crew could help him recover from a transporter accident that had split his personality in two. No reasonable jury anywhere in the galaxy would find Spock guilty of aggression.

Finally, what's the jury going to say about Worf's love bites? Worf and his mate are like Klingons everywhere. Their romantic attraction for each other is expressed in typical Klingon fashion, with lots of rough stuff and mutual biting with their jagged Klingon teeth. Broken bones, wounds, and blood are part of Klingon-style joy of sex. Consenting adults of this species find the intense physical stimulation pleasurable, and they wear their love wounds as badges of honor. Klingon sexual practices may seem raucous, perhaps even kinky, but they are not aggression. The participants willingly—eagerly—accept the possible harm and injury that comes with such foreplay.

The lesson from these four incidents? Aggression comes in different packages, ranging from potentially lethal violence to nonlethal verbal threats. And in reaching a verdict about aggression, we have to consider intent. These incidents also suggest another aspect of aggression: different individuals have their own, unique ways of expressing aggression.

What Would You Do?

Psychologists often ask people to imagine how they would react to various hypothetical situations. This is not idle curiosity: their aim is to find patterns in the answers that different people give. Such studies have proved what everyday observations only hint at: "aggression" is not a single characteristic that a person either has or doesn't have. Instead, different people exhibit different forms of aggression in varying degrees. Some people, for example, are aggressive verbally (including "flaming" in e-mail messages), but rarely express that aggression physically. Others show exactly the opposite pattern of behavior—they are verbally passive, but prone to outbursts of physical aggression.

Several different forms of aggression are illustrated by the following statements from a widely used questionnaire.[5] Read each of them and select the one that best applies to you. Be as truthful as you dare to be—and, of course, do your own work.

1. Given enough provocation, I may hit another person.
2. I can't help getting into arguments when people disagree with me.
3. Some of my friends think I'm a hothead.
4. I am suspicious of overly friendly strangers.
5. None of the above applies to me.

Now that you've made your selection, compare it with those you'd expect from several of *Star Trek's* familiar characters. Start with statement 5. The one Starfleet officer who could truthfully select this statement is Deanna Troi, ship's counselor aboard the *Enterprise*-D. She's as tranquil, unflappable, and nonaggressive as any humanoid imaginable. About the only time her calm demeanor breaks down is when her mother, Lwaxana Troi, gets under her skin, and even then, Deanna's irritation is mild by human standards. It's not her Betazoid genes that are responsible for her placid nature, for other Betazoids—including Deanna's mother—do fly off the handle from time to time. And it's not that Troi is incapable of emotional reactions, because she experiences intense sensual pleasure when eating a good piece of chocolate (TNG-5, "The Game") or when mothering her child (TNG-2, "The Child"). And for all her calm, the counselor does have it within her power to be violently aggressive. When she's jealous, she murders Lieutenant Worf by firing a fully charged phaser into his chest (TNG-7, "Eye of the Beholder"). Luckily for Worf, the murder happens only in her imagination; she certainly doesn't act on her anger. How Troi manages to muzzle her aggressive feelings and behaviors is a mystery. Whatever the explanation, she would definitely select statement 5 from the questionnaire.

Troi's calm, even temperament couldn't be more different from the hard-edged, fly-off-the-handle disposition of Leonard McCoy, chief physician on board the original *Enterprise*. McCoy routinely insults and belittles the ship's first officer, Spock, whose lack of emotion galls McCoy. Ignoring protocol, McCoy also speaks harshly to the ship's captain, James Kirk, bluntly criticizing his judgment. But much as McCoy mouths off, he rarely hits anyone—his aggression is more bark than bite. Assuming he's honest with himself, McCoy would probably select statement 2 as the best description of his tendencies.

Worf, the Klingon security chief on the *Enterprise*-D, seems to be just the opposite of McCoy. Worf doesn't need abusive words to vent his spleen—he does a pretty good job of holding his tongue in the line of duty. But when there's sufficient provocation, Worf can explode into physical violence. Being honest with himself, Worf would agree that statement 1 is the most apt for him. Unlike McCoy's, Worf's bite *is* much worse than his bark (especially with those Klingon teeth).

Voyager's chief engineer, B'Elanna Torres, is different from both McCoy and Worf—she bites as well as barks. Among Starfleet crew members, Torres stands out as one of the most fearless and outspoken officers. Statement 3 describes her perfectly. Perhaps her combination of aggressive tendencies can be traced to her hybrid biology. Being part human, she has the verbal aggressiveness of McCoy, and being part Klingon, she has the physical aggressiveness of Worf. But that's just a guess. In any event, Torres is an ideal companion on a dangerous away mission.

Now consider statement 4. Which Starfleet personnel would fit this description of mistrust and suspicion? We'd pick Constable Odo, Deep Space Nine's shape-shifter security chief. He seldom gets angry, and he rarely tells anyone off. So statements 1 and 2 don't seem to fit. When someone offends him, Odo is like Worf—he usually keeps it to himself. However, Odo becomes very suspicious—sometimes downright hostile—when strangers are friendly, especially when the friendliness radiates from a Ferengi or a Cardassian. Odo knows that his unsmiling face and penetrating eyes contribute to his effectiveness as an enforcer of laws. He tells one miscreant, "I pay special attention to my scowl. An air of suspicion is important in my line of work" (DS9-5, "A Simple Investigation"). Odo is the very model of the form of aggression captured by statement 4. His aggression needs no fists, no knives, not even threatening words: his dirty looks do the job perfectly well.

This little test makes it clear that different *Star Trek* humanoids typify different styles of aggression. The same is true of twentieth-century humans. Thousands of them, from various backgrounds and socioeconomic groups, took an expanded version of the test you took. Their responses demonstrated that different people tend to rely on different forms of aggression. We can label these forms *physical aggression* (illustrated by statement 1), *verbal aggression* (statement 2), *anger* (statement 3), and *hostility* (statement 4). These four forms of aggression are, of course, related, and unlike Worf, Odo, McCoy, and B'Elanna, many humans don't fit neatly into just one category or another.

Watch What We Do, Not What We Say

Questionnaires are just one way to study aggression. Another way is to measure the physiological responses of people at the very moment they are feeling aggressive. Such measurements reveal that heart rate, blood pressure, and levels of certain hormones all rise, providing clear physiological signs of anger and arousal. Imagine, though, a species so different from *Homo sapiens* that it has never experienced aggression or anger. How on Earth—or anywhere else, for that matter—could a member of that species figure out what the rest of us mean when we say things like "You make my blood boil" or "I'd like to beat your brains out"?

The Iyaarans face precisely this dilemma (TNG-7, "Liaisons"). They've studied video records of other species, including *Homo sapiens*. But all their study hasn't helped. They have no clue what love is, and they don't get the idea of pleasure. More to the point here, the Iyaarans don't understand aggression.[6] To remedy their ignorance, they devise an interesting study: under carefully controlled conditions, they try to provoke aggression in one of the *Enterprise*-D's crew members. Ambassador Byleth, the Iyaaran visitor who carries out the experiment, selects Lieutenant Worf as his unwitting guinea pig.

During dinner in Ten-Forward, Byleth launches an escalating barrage of intentional provocations. He begins with a brusque announcement that the food is unacceptable. When Worf apologizes, Byleth orders him to bring a new plate of food with higher protein and enzymatic value. Dr. Crusher politely points out that service is buffet style, inviting Byleth to serve himself. But Byleth again barks at Worf, "Bring me new food." In grudging recognition of his duty as Byleth's host, Worf accedes.

Worf's mouth says "I don't mind," but his behavior tells a different story. Grabbing a huge knife, Worf proceeds to stab and hack at a large slab of roast meat on the buffet table. To most humans or Klingons, the fury in Worf's actions would make it clear that he's not thinking about carving a portion of meat to serve to Byleth. He's thinking about carving up Byleth. But because he's from way out of town, Byleth is oblivious to the simmering aggression that fuels Worf's behavior.

Incidentally, this kind of behavior is so common on our planet that it has a name of its own: displaced aggression. This term refers to aggressive behavior directed at a person or object other than the one that instigated the aggressive feelings. Worf seethes at Byleth's rudeness, but his role as diplomatic host keeps him from retaliating. So he vents, or displaces, his anger on the helpless piece of roast meat. This form of aggression is common in everyday life. Imagine a small child tormented by his big brother. The child is angry, but is afraid to take on his older sibling. So he turns his anger on a younger, smaller, defenseless sister, or one of the family pets. Perhaps you've been chewed out by your supervisor or teacher, but couldn't talk back to defend yourself. Instead, you find yourself flying off the handle at an innocent co-worker or fellow student. Your aggression is displaced from its true source to a surrogate.

Returning to Byleth's study of aggression, his experiment reaches a surprising, but successful, conclusion. A poker game is arranged in the hope that this shared activity will calm the roiling diplomatic waters. But Byleth's interest is less in winning the pot than in stirring it up. He provokes Worf with constant, blatant cheating during the game. Finally Worf explodes, and the accusations and counter-accusations ratchet up to what diplomats call *causus belli*, an excuse for all-out war. On his way toward mincing the ambassador, Worf doesn't mince words:

> **WORF:** You are an insulting, pompous fool. And if you were not an ambassador I would disembowel you right here.
>
> **BYLETH:** [thrusting his finger repeatedly into Worf's chest] Don't let my title inhibit you, Klingon.

To put it mildly, Worf and the ambassador come to blows. It's one of the best *Star Trek* fights ever. They whack, kick, and punch each other with considerable gusto. Each time Worf smashes him or throws him across the room, Byleth expresses sincere gratitude, not anger.

Finally, Worf head-butts the ambassador with a bone-crunching impact that would make a National Football League tackler envious (Klingon skulls are *very* hard). The ambassador returns the favor by telling his Klingon opponent: "Wonderful. Very good. Thank you, Lieutenant Worf. I think I understand now. That was a very effective demonstration. . . . If you'll excuse me, I'd like to document this experience." And Byleth goes off to write up the results of his experiment, leaving an astonished Worf in his wake.

For us human beings, aggression is so common that it's hard to understand what the ambassador could have learned from this experiment. What is there that he couldn't have picked up from video footage of a bloody war, or of the fistfights that occasionally break out in our world's deliberative bodies? The answer, to use Byleth's own word, is experience. In his escalating feud with Worf, Byleth may finally have figured out what aggressive urges *feel* like.

Byleth should have learned another lesson, one that humans would be wise to take to heart: one person's attack often begets another's counterattack. Aggression breeds more aggression. And, disturbingly, nowhere is this truer—or more dangerous—than in the arena of domestic violence.[7] Bitter words spawn bitterer words, throwing more fuel on the fire. As studies of intrafamily fighting have confirmed, verbal aggression and physical aggression go hand in hand—arguments can easily escalate to blows, or worse.[8]

But isn't it best to get things off your chest rather than keeping your anger pent up inside? Why shouldn't a wife scold her spouse when he leaves the toilet seat up after he's finished urinating? Isn't it healthy for her to speak her mind? Not necessarily. Outspoken criticism can have unintended consequences. Verbalizing a complaint, especially against one's significant other, seems to trigger thoughts of additional complaints—griping about the toilet seat is likely to bring to mind other annoyances such as the spouse's lazy habit of leaving his shoes in the front hallway. These associations occur naturally because the human mind uses a particularly powerful and flexible form of memory in which stored information is linked to other stored information. A by-product of this excellent memory is the fact that one hostile thought can recruit other hostile thoughts to which it is linked. Once the associations are triggered, you begin to experience a cascade of grievances. As a result, your anger escalates.

By itself, expressing a grievance isn't inevitably destructive, particularly if you voice it in a calm, nonaccusatory fashion and if you stick to the issue at hand. But you also need to gauge the mood of the person you're about to criticize. For criticism to be effective, that person needs to hear your complaints as constructive problem-solving, not personal attacks. Angry listeners aren't good listeners. Unfortunately, this sound advice is hard to follow when push comes to shove. All too often, a litany of wrongs is recited in accusatory tones of frustration and anger. Consequently, those messages fall upon defensive ears. And what happens then? Anger builds, and anger can be prelude to hostility. To quote Benjamin Franklin, "Whate'er's begun in anger, ends in shame."[9] If you're already riled up and ready to take offense, it's hard to remain calm and dispassionate when you're being peppered with critical, threatening words.

This important realization suggests how to be constructive when you have no choice but to be critical. First and foremost, don't confront someone with a complaint during the heat of your anger. Let your temperature drop below the boiling point before broaching the subject. Thomas Jefferson recommended counting to ten before speaking when angry, and counting to one hundred when *very* angry—both excellent ideas. And when you do finally verbalize that criticism, don't present the complaint as a challenge to the other person's character or genetic background or intelligence. Instead, try to frame it as an idea about how to satisfy one of your needs, irrational though it may be.

Letting Off Steam

People often draw an analogy between pent-up aggression and the buildup of pressure within a closed container. Imagine that aggressive feelings are like the gaseous by-products of an oil refinery.

If these gases are confined within a tight enclosure, pressure can build up, resulting in a devastating explosion. To prevent such explosions, oil refineries carefully vent the gases, recycling them or simply allowing them to burn off. If it behaved like a gas, pent-up aggression could be explosive, too. If you buy this metaphor, the solution would be to vent your aggressive feelings a little at a time before they build up to dangerous levels.

Sigmund Freud was one person who subscribed to this "ventilation" view of aggression. He suggested that humans rely on several different kinds of "release valves" for pent-up aggression. Humor is one. Freud said that when you crack a joke or play a prank, you're diffusing aggressive energy that would otherwise culminate in violent aggression.

How seriously should we take this claim that aggressive impulses power humor's engine? Political humor, which often has a nasty edge, seems to fit this theory. So does raw humor that denigrates ethnic groups or mocks the foibles of individuals. Freud could have pointed to Lieutenant Commander Data as living proof of his theory about the link between aggression and humor. Ignoring the fact that Data is an android rather than a biological entity, his underdeveloped sense of humor is consistent with his nonaggressive behavior. As we saw in Chapter 2, Data is puzzled by humor: for the longest time, he fails to understand the jokes cracked by other crew members. He's so troubled by his incomprehension of humor that he takes lessons from a computer-generated twentieth-century comedian (TNG-2, "The Outrageous Okona").[10] But in the joke-telling department, Data is an utter dud. Freud could readily "explain" Data's poor sense of humor: because he's nonaggressive, Data has no need for humor, since humor's main purpose is to discharge impulses Data doesn't possess in the first place.

Freud's theory runs into trouble, however, when it comes to Vulcans. From the original *Enterprise's* Spock, who is just half Vulcan, to *Voyager's* full-blooded Vulcans, Tuvok and Vorek, members of that species epitomize humorlessness. To put it nicely, a Vulcan joke book would be an extremely slim volume. Expressionless, unsmiling faces are a Vulcan trademark. For example, when someone teases Julian Bashir (a genetically enhanced human) by saying that his mathematical wizardry makes him sound like a Vulcan, Bashir comes back with the perfect defense: "If I'm a Vulcan, how do you explain my boyish smile?" (DS9-6, "A Time to Stand").

The fact that Vulcans have no use for humor, however, does not mean that they are therefore free of aggression. On the contrary,

Vulcans are fully capable of feeling anger (as Chapter 2 documents). It's just that they have learned to control their emotions, including anger. Because they have all that bottled-up aggression, Vulcans should be expected to discharge it with a joke every now and again. But, sorry, Dr. Freud, Vulcan jokes are rare.

So it's probably wrong to think of aggression as a kind of explosive gas that builds to dangerous levels. It's not a choice between hitting your bratty brother or making a nasty joke about him. Instead, aggression results from a preexisting blend of ingredients ready to be ignited when the right environmental catalysts are encountered. But this brings us back to our earlier question: Why would nature create this explosive mixture and then put some of its ingredients into each of us? To dig up the answer, let's turn to the mining excavations on a remote Alpha Quadrant planet.

The Rhyme and Reason of Aggression

For fifty years, the underground caves on Janus VI have been a rich, dependable source of minerals for the United Federation of Planets. But now that source is threatened. When a new level is opened in Janus VI's pergium mine, miners start dying one by one, each incinerated to a small pile of burnt cinders. The death toll is fifty and rising—fast. The Federation is worried because it needs pergium to power its life-support systems (TOS-1, "Devil in the Dark").

When Captain Kirk leads a investigating mission to Janus VI, he discovers that the perpetrator is not exactly someone you'd find in a police lineup on Earth. The killer is a huge centipedelike creature, called a Horta, that makes its home in the pergium mine. The Horta's skin secretes a corrosive acid that can burn through the mine's rock walls like a hot knife through butter. In a flash, that acid will make a miner a crispy critter. No question about it: The Horta is one dangerous creature.

With considerable luck, Kirk and Spock manage to wound the Horta. Once they have the beast cornered, you'd think the next step would be obvious. The Horta's attacks on the miners would seem to leave no doubt that this ugly, wanton creature has to go. However, one of *Star Trek's* many delights is that things are not always what they seem.

Spock mind-melds with the creature, and learns that this Horta is the only remaining individual of its kind. Every 50,000 years, Horta eggs hatch and produce the new creatures that will replenish the species. The Horta that Spock and Kirk have wounded is supposed to guard those precious eggs, which happen to be incubating in the newly opened section of the mine. Now things make sense: The Horta killed in order to protect all the unborn of its species. Its behavior was not random, unprovoked violence, but rather an understandable reaction when faced with a mortal threat to its progeny. Kirk and Spock broker a deal: The Horta and its offspring will live unmolested in certain sections of the mine and, in turn, the miners can tend to the pergium business without fear of being turned into charcoal briquettes.

Star Trek does not flinch from showing us the dark side of aggression,[11] but it also shows that when conditions are right, aggressive behavior may be justified by its positive contribution to our survival. *Star Trek* suggests that nature's creatures—*Homo sapiens* included—don't have a general instinct for aggression. However violent our kind seems to be, we may not be naturally violent creatures who crave violence for violence's sake. Instead, nature has built into us an instinct to survive, and aggression provides one means of securing and defending the resources we need for survival.

Our hominid ancestors probably developed their aggressive behaviors over the millions of years that elapsed between our African genesis and today. Remember that aggression, like any other behavioral adaptation, is not the product of forethought or of some grand plan. Nature stumbles onto behavioral adaptations as a result of genetic "accidents" that produce new physical traits, which enable new behavioral abilities. There's no purpose or plan, other than obedience to a single overriding principle: That which promotes the promulgation of one's genes will tend to proliferate in the gene pool. Keeping this in mind, we can make some guesses about the factors that encouraged the evolution of aggressive behavior in our ancestors.

A good defense is the best offense. Once our primate ancestors abandoned the safety of the treetops for the more exposed ground below, they had to worry about becoming snack food for some faster, stronger predator. To defend themselves, individual hominids gathered in groups, the first hominid embodiment of the "strength in numbers" axiom. Fossil evidence indicates that males were the mainstay of these defensive alliances. Contemporary field research into the behavior of living social primates supports the view that males provide the group's frontline defense against attack. This defensive role explains, in part, why primate males are generally larger and more heavily muscled than their female counterparts.

And how did these individuals behave when threatened? If gorillas and chimpanzees are any clue, our ancestors stood on their hind legs (making themselves look as big as possible) while baring their teeth and shrieking at the top of their lungs (universal signals of intense anger). They also had a look in their eyes that signaled ferocity, the same look that one Massachusetts driver shoots at another in rush-hour traffic. For Klingons, too, the eyes are especially powerful messengers of aggression, as Worf explains to a young admirer (DS9-5, "Children of Time"):

> **YOUNG BOY:** Are you the son of Mogh?
>
> **WORF:** Yes.
>
> **BOY:** Is it true you can kill someone just by looking at them?
>
> **WORF:** Only when I am angry.

Our primate relatives' defensive warning signals were undoubtedly accompanied by intense physiological arousal, a vigorous flow of blood to the brain and the muscles of the arms and legs, accelerated breathing, and profuse sweating. These reactions are all caused by a rush of adrenaline stimulated by the autonomic nervous system. Accompanying these strong physiological reactions is a strong sense of anger and fear. Our ancestors hoped, of course, that their defensive show of anger would convince a threatening predator to turn and flee. Their shows of defiance were spectacles designed to cow an opponent and secure victory without shedding a drop of blood. Sometimes the theatrics worked; sometimes they didn't. If their warning signals were ignored by the predator, our ancestors had two options: flee or fight.

Make mine medium rare. While they were adapting to life on the ground, our ancestors acquired a taste for meat. And this appetite demanded new hunting skills. Our ancestors—again mainly the males, because of their size and strength—became predators themselves. Teams of males went hunting, coordinating their attack by means of crude gestures and perhaps occasional grunts. And because hunting was a social act, the fruits of the kill were shared. The result was the evolution of greater social cohesion. Indeed, some believe that the roots of human language can be traced back to these bands of cooperating primates, whose utterances and gestures increased the potential for social exchange. And it wasn't just the males who exploited communication. Primate females formed close social connections, actively participating in the rearing of one another's young. They probably relied heavily on vocalizations, with different calls signalling different threats.

I saw that parking space first. Anthropologist Richard Leakey looks back no further than 10,000 years to pinpoint the most important circumstance promoting human aggression: he blames the farmers.[12] As hunter/gatherers tens of thousands of years ago, our ancestors had to cooperate to survive, and they lived in groups that tended to move about. Ownership of land for any extended period was probably unheard of. But approximately 10,000 years ago, groups of people began to settle and raise crops. Consequently, territory became important. Resources were concentrated, becoming more tempting to others. With the development of agriculture came the birth of commerce and materialism, and these spawned greed and envy. According to Leakey, these conditions spawned human aggression.[13]

While greed and envy are not characteristics to be proud of, there is a silver lining to this cloud. Leakey's theory lays the blame for aggression on social conditions, not hereditary dispositions imprinted in our brains. Social conditions can be altered within a lifetime or two, whereas biologically ingrained drives take generations to modify.

Wherever our aggressiveness comes from, everyone agrees that our enlightened species ought to be able to curb its anger and violence. One way we can modify aggression's violent consequences is to express it in effective but nonlethal words and deeds. In fact, we engage in this kind of ritualized aggression all the time, without realizing it.

Rituals of Aggression and Dominance

Every single one of the male chimpanzees is very agitated. Something has stirred the emotional waters, and Jane Goodall has captured it all on film. There, in Tanzania's Gombe Stream Reserve, the wild chimpanzees glare at one another, playing chicken with their eyes, each daring the other to break eye contact first. And they are noisy about it. They stamp their feet loudly and slap their sides, snarling at one another through open lips and bared teeth.

They're like human soccer players who have just won their game. The chimps tip their heads backward, thrust their chests outward, and raise their arms over their heads in what looks like a display of triumph.

Among the humanoid species in the *Star Trek* universe, none embrace aggression more than the Klingons. They pride themselves on being warriors whose traditions begin and end with honor, loyalty, battle, and death.[14] How can a fierce race such as the Klingons avoid internecine battles? How do they keep from killing one another? Klingons have clearly defined social hierarchies. Everyone recognizes the lines of authority, and individuals rarely challenge them. When male Klingons congregate in the Hall of Warriors on their home planet, Qo'noS, their behavior is reminiscent of what Goodall saw in Tanzania (DS9-5, "Apocalypse"). Like Goodall's chimpanzees and other Terran primates, Klingons engage in ritualized confrontations designed to sort out their dominance hierarchy (the pecking order, as it's sometimes called). These rituals establish who's top Klingon, without the mess and carnage of lethal violence. Klingons fight among themselves mainly to defend family honor (TNG-4/5, "Redemption," I, II) or as part of a struggle for political power (TNG-4, "Reunion").

In species after species, when members live in groups, social organization emerges—the *pecking order* is clear. The term "pecking order" itself describes the procedure that many birds, domestic hens included, use to establish their social order. Within a flock of hens, one member—the alpha hen—dominates all the others, and her dominance is expressed by her freedom to peck all the others without fear of being pecked in return. Immediately below this alpha hen is another bird—naturally called the beta hen—who enjoys the privilege of pecking all the others except the alpha hen. And so on down the line. The underling birds don't insist on an eye for an eye, a peck for a peck. When an underling is pecked, it does not peck back—that would be a very bad idea. Instead, the peckee accepts the peck as a reminder of its standing within the flock. Naturally, the alpha hen is entitled to a number of perks. When the corn buffet is served, the alpha hen gets to eat first, and when it's time to sleep, the alpha hen has the pick of prime roosting spots. A hen that is newly introduced into the flock has to earn her place in the hierarchy, working her way through the pecking gauntlet.

Pecking, of course, is peculiar to birds; other species employ their own unique expressions of dominance tailored to their own makeup. Among a wide variety of social animals, one overriding principle emerges: these dominance battles rarely end in death.[15] Species with potentially lethal weapons—claws, horns, poisonous fangs—rarely use them to injure an opponent during dominance battles. They reserve those weapons for defense against predators or for acquiring food.

Baboons and some other primates have developed interesting refinements of the social dominance hierarchy. Leadership is not exercised by a single individual, but rather is shared among several high-ranking males. When the group is looking for food, this coalition leads the hunt. When it's time to enjoy the fruits of the hunt, the entire leadership coalition gets first dibs. Social climbing is another interesting twist in primate social organization. This option allows an individual with low status to "leapfrog" his way up the social ladder, bypassing several rungs. All the social climber has to do is befriend one of the leaders, who, in turn, will back the social climber in confrontations. "Befriending" in monkeys involves grooming the leader or securing a tasty tidbit of food for him. The parallels in human social organization are abundantly obvious.

Whenever you see a collection of Klingons, there's no doubt who is in charge. They care little about feel-good strategies designed to build consensus and promote bonding among group members. Instead, the leader barks orders to his or her underlings, much as the dominant, alpha male does in a chimpanzee or gorilla group on Earth. Klingons don't have a monopoly on social dominance hierarchies, however. Even within the egalitarian "societies" of the *Enterprise*-D and Deep Space Nine, we find an "alpha" leader whose position atop the dominance hierarchy is made clear by symbols of rank. Gracing the captain's uniform is that small row of gold "pips" denoting Starfleet rank. The captain's special chair occupies center stage on the bridge, and no one sits in it without explicit permission.[16] Salutes have been dispensed with, but the crew still addresses the captain by title. The captain routinely uses the phrase "my ship," not "our ship." Moreover, the captain does not tolerate interruptions while speaking, as young Wesley Crusher learned in "Where No One Has Gone Before" (TNG-1).

The development of dominance hierarchies on board *Star Trek* vessels can be seen in some very early episodes of *Star Trek: The Next Generation* and *Star Trek: Deep Space Nine*. As in all groups and societies, the newly assembled crew members' first interactions set up the hierarchy and establish patterns for their subsequent interactions. Look at what happens when Commander William Riker first comes on board the *Enterprise*-D, where he is to serve as first officer (TNG-1, "Encounter at Farpoint"). After beaming aboard, Riker is escorted by Lieutenant Yar to the command deck and formally introduced to Captain Picard. Seated in the captain's chair, Picard ignores Riker's presence for an uncomfortably long time, and then neither stands nor offers his hand in welcome. This brusque treatment continues in the captain's ready room, where Picard's behavior again shows who is boss. It isn't until Riker successfully completes a complicated docking maneuver that Picard acknowledges his competence to join the "team."[17]

The dominance hierarchy also has to be sorted out on board Deep Space Nine. When Benjamin Sisko arrives on the space station, he is immediately at odds with Major Kira Nerys, a Bajoran female assigned to be his second-in-command (DS9-1, "Emissary"). She and others in the Bajoran resistance have struggled for decades to free their home planet from Cardassian occupation. After finally driving out the Cardassians, the Bajoran provisional government invites Starfleet's help in overseeing Deep Space Nine, a space station abandoned by the Cardassians. Starfleet accepts the invitation. However, many of the former resistance fighters—including Major Kira—are cynical about Starfleet's motives. They suspect that one tyranny is about to fill the void left by the departure of another. In her characteristically blunt style, Major Kira voices her resentment of Starfleet's presence. Sisko and Kira engage in several shouting matches before the captain manages to establish his authority over her and win her confidence. Their hierarchical relationship is established, but a comfortable working relationship takes longer to build. As often happens in groups and societies here on Earth, an external threat provides the glue that bonds them into an effective team.

Murder: It's Killing Us

When we were growing up, they used to tell us humanity had evolved; and mankind had outgrown hate and rage. But when it came down to it, when I had the chance to show that no matter what anybody did to me I was still an evolved human being, I failed. I repaid kindness with blood. I was no better than animal."

Miles O'Brien (DS9-4, "Hard Time")

Humans can be very violent. But—resorting to the "everybody does it" justification favored by human children and politicians—other species are violent, too. In fact, some species outdo us. Considering just mammals currently known on Earth, hyenas, monkeys, and lions all have higher per capita intraspecific murder rates than our own species.[18] However, we can take pride in being the only Terran species to have developed weapons capable of killing on a grand, wholesale scale. Even on our best, most pacific day, humans have a serious problem with violence.

All sorts of professions have tried to explain why humans kill other humans. Biologists, social scientists, legal scholars, philosophers, and theologians all have put in their two cents' worth. But frankly, no one has a lock on the truth. There may well be several equally valid answers.

One interesting possibility is that our species is not inherently bad or dangerous, just ignorant. In particular, we're ignorant of how to use limited ritual aggression, that life-saving strategy that other species are so expert in.[19] When humans get into a fight with one another, the fight can easily escalate until one combatant is dead. And there may be good reason for our inability to quit fighting before violence gets out of hand. Our hominid ancestors didn't have the deadly weapons that other mammals have. Look at yourself in the mirror: No claws, no beak, no long canine teeth for ripping and tearing flesh; no poisonous fangs or glands that secrete lethal chemicals. And because these lethal specializations were not part of their arsenal, our hominid ancestors did not have to develop rituals for controlling intraspecific violence. They had no reason to keep fighting in check, because fighting so rarely led to death.

But at some later point in our history, a few of our ancestors made a great and disastrous discovery: they learned how to construct and use weapons. Pointed sticks and sharpened stones gave them a real edge over their unenlightened, unarmed competitors. At this stage, our hominid ancestors became like other species: they could kill one another with ease. Unfortunately, the brains behind their deadly weapons were still ignorant of the rules of ritualized fighting. Thus we kill one another because we've never learned to rechannel our aggression into ritual combat, like Goodall's chimpanzees and *Star Trek's* Klingons.

Writing in the early 1960s, the Austrian ethologist Konrad Lorenz was optimistic that the same brain that learned to build weapons could learn how to control their use.[20] Looking several centuries into the *Star Trek* future, it seems that Lorenz was wildly overoptimistic. During the twenty-first century, Earth suffered a devastating world war whose nuclear fallout seriously compromised the planet's ecosystem. This pivotal event led Earth's humans to abandon warfare as a means of settling differences, forming a united world government at the beginning of the twenty-second century (*Star Trek VIII: First Contact*). But by the middle of the twenty-second century, Earth was again at war, this time with the Romulan Star Empire (TOS-1, "Balance of Terror"). Earth's forces dealt the Romulans a solid defeat, and a demilitarized zone was established in space separating the Romulan planets from the rest of the galaxy. Soon afterward, in 2161, the United Federation of Planets was formed, along with an agency to oversee exploration, diplomacy, and defense. That agency, of course, was Starfleet. But tensions remain, alliances shift, new enemies emerge, and wars are still endemic.

If *Star Trek* is indicative of what's to come, several more centuries of cultural evolution will be insufficient to eliminate some of aggression's roots. And perhaps those resistant roots are to be found within the humanoid brain, an organ unlikely to change much in only a couple of hundred years.

Raging Hormones

Commander Riker and his estranged father, Kyle, have not seen each other in 15 years. They are using the *Enterprise*-D's gymnasium for a grudge match of anbo-jytsu, a martial art.[21] Katherine Pulaski, the ship's doctor, is an old girlfriend of Kyle Riker's, and she and Deanna Troi, the ship's counselor, are comparing notes (TNG-2, "The Icarus Factor"):

> PULASKI: I'm just glad that humans have progressed beyond the need for barbaric display.
>
> TROI: Have they? Commander Riker and his father are in the gymnasium, about to engage in a barbarism of their own.
>
> PULASKI: Don't remind me. It's something of which I do not approve.
>
> TROI: In spite of evolution there are still some traits that are endemic to gender.
>
> PULASKI: You think they're going to knock each other's brains out 'cause they're men?

What makes males so eager to fight? The usual answer—"Men are pigs"—is not precise enough to be meaningful. The answer "Men are angrier than women" doesn't work either. Results from the aggression questionnaire described earlier in this chapter show that males and females are equally angry. But this anger tends to find different modes of expression in the two sexes. Males rate themselves, and are rated by others, as much higher on physical aggression. As Dr. Pulaski put it, they "knock each other's brains out 'cause they're men."

On average, human males tend to be bigger and more heavily muscled than human females.[22] These differences in body structure can be traced back to the time of conception, when genetics determines whether the embryo will blossom into a male or a female. The XY chromosome pairing that specifies "male" triggers the production of different types and levels of hormones in the embryo than does the XX chromosome pairing that specifies "female." These different histories of hormone exposure continue throughout life.[23]

For humans, sexual differentiation really thunders into view during puberty. In males, testosterone levels go through the roof, triggering the transition from boyhood to manhood. The male's voice drops, and he experiences a growth spurt. And as testosterone levels rise, aggressive behavior increases. So does the crime rate, which reaches a maximum among males thirteen to seventeen years old. Study after study has shown that increased aggression goes hand in glove with high levels of testosterone. In fact, "[t]he evidence is incontrovertible that the male brain pattern is tuned for potential aggression; that the action of male hormones acting upon a predisposed male brain network is the root of aggression."[24] When aggressiveness reaches dangerous levels, the symptoms can be squelched by treatment with estrogen or, if all else fails, by castration (which removes the glands responsible for the secretion of testosterone). By castration, "a wild stallion is transformed into a docile horse, a savage bull becomes a plodding ox, and a rowdy dog is turned into a sedate pet."[25]

A Lesson from the Vulcans

Vulcans are renowned for their dispassionate, logical personalities. In their vocabulary, the phrase "not logical" is unsurpassed as a pejorative. But they weren't always that way, and their transformation may hold out some lessons for our kind.

Several thousand years ago the Vulcans were highly emotional and extremely violent. War was common, and murder was the routine way of securing a mate. These destructive behaviors jeopardized the Vulcans' very existence. Fortunately, the Vulcan philosopher Surak convinced his people to renounce emotions. They were, ever after, to govern their lives strictly on the basis of logic (TOS-3, "The Savage Curtain"). Unfortunately, the record tells us nothing about how Surak sold this idea to the Vulcans. How did he manage to accomplish what Jesus, Gandhi, and Martin Luther King Jr. tried but failed to do on Earth: teach people to live together in harmony and peace?

Zero population growth. One possibility is that Surak encouraged steps that would minimize crowding. Intraspecific violence increases dramatically in crowded living conditions. There are a couple of reasons why crowding has this effect. First, resources—food, water, shelter—become scarce, intensifying competition between haves and have-nots. Second, crowding promotes depersonalization—we are more likely to lash out aggressively against others we don't know (think about your own impatience and rudeness while driving to work in a traffic jam).

There are two ways to minimize crowding: you can spread the existing population as widely and evenly as possible, or you can control population growth. These strategies are often seen in animal species, in which predators keep the population down and migration keeps any one locale's resources from being depleted. Perhaps Surak realized that crowding contributes to aggression and convinced the Vulcans to take appropriate measures.

Zero tolerance for crime. Or maybe the Vulcans followed the lead of the Edo, those sensual humanoids who inhabit Rubicun III, where the sun always shines, the days and the people are beautiful and friendly, and crime of any kind is just about unknown. The Edo achieved this last component of their idyllic state by following one simple rule: Any crime is quickly punished by death, no ifs, ands, or buts (TNG-1, "Justice"). And when Edo law says that *any* infraction is a capital offense, it means exactly that, as young Wesley Crusher discovers when he accidentally steps on the grass while chasing a ball. Stepping on the grass is an infraction, so Wesley is immediately sentenced to death. No wonder people on Rubicun III are so law-abiding.

That's using your lobes. There's one other, more likely, explanation for the Vulcans' success in taming violence: they learned to use the frontal lobes of their brains more effectively.

The illustration below sketches the brain pathways that initiate aggressive behavior. Notice first that the hypothalamus occupies a place of honor at the diagram's center. Electrical stimulation applied to strategic areas of the hypothalamus evokes rage. This is the brain region that Walter Hess stimulated in his experiments on feline rage seventy years ago. It's also the region where the Aquitarians inserted stimulators to trigger rage in their humanoid prisoners. The hypothalamus exerts a major influence on the autonomic nervous system, which controls heart rate, temperature, blood pressure, and respiration. In addition, the hypothalamus stimulates the pituitary gland, causing the release of hormones into the bloodstream. These hormones, in turn, accelerate metabolism, making increased energy available to the body. All of this conspires to make it easier to act out your rage and anger.

So the hypothalamus does some very important things, but it has lots of help. Overseeing the operation of the hypothalamus are components of the limbic system, notably the amygdala. These structures, in turn, are interconnected with regions of the frontal lobes. Together these regions form a feedback circuit in which the frontal

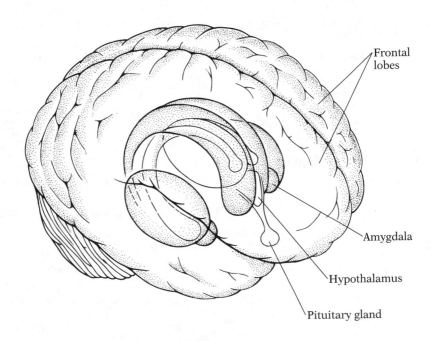

lobes tend to keep the limbic system and hypothalamus on a short leash. When this system is let off the leash, there can be trouble.

In humans, damage to certain areas of the frontal lobes can snap the leash, producing unprovoked aggressive outbursts. The amazing thing about these outbursts is not how violent they are, but how reckless the behavior seems. Normally, the frontal lobes put the brakes on potentially dangerous or risky actions. The removal of their inhibitory influence is described as a release from inhibition or, slightly more euphoniously, *disinhibition*. One common cause of frontal lobe disinhibition is trauma—a penetrating wound or a severe blow to the head. After such an injury, a person may exhibit signs of temporal myopia, or blindness to the long-term consequences of his actions. And this condition persists.[26]

Frontal lobe disinhibition can also come from drugs, such as cocaine, amphetamines (in large doses), or alcohol. With alcohol in particular, many scientists believe that drug-induced temporal myopia is to blame for the drug's tendency to increase violence.[27] According to this view, alcohol limits your ability to appreciate the future consequences of your actions. And this nearsightedness promotes recklessness.

Compared with the drinking establishments of twentieth-century Earth, let alone those of the eighteenth or nineteenth centuries, drinking establishments on starships experience remarkably few instances of drug-induced aggression. Despite a few well-known exceptions, barroom brawls are a rarity.[28] Ten-Forward, the *Enterprise*-D's lounge, and Quark's, the watering hole on Deep Space Nine, are oases of calm and tranquility. Given how much drinking goes on in these establishments, there must be some reason why there's so little drug-induced temporal myopia, and consequently so little aggression. One possible reason is that in both lounges, most of the mixed drinks contain synthehol, an alcohol substitute invented by the Ferengi. Synthehol relaxes, but produces no hangover. More importantly, synthehol seems to leave the drinker's frontal lobes relatively unaffected.

So we know that the brain areas involved in the instigation of aggressive behavior are kept in check by neural signals from the frontal lobes.[29] When Jean-Luc Picard manages to hold his tongue during one of Q's insulting diatribes, it's Picard's frontal lobes that deserve his eternal gratitude. When James Kirk holds his phaser fire and doesn't kill the Gorn, the hideous looking creature owes its life to Kirk's frontal lobes (TOS-1, "Arena"). The same goes for you, too. Think about this the next time you're stuck in rush-hour traffic and your impatience is building. When you feel the urge to honk your horn or cut into another lane, but instead just breathe deeply and count to ten (as Thomas Jefferson suggested), your sensible self-restraint is a gift from your frontal lobes, and one that protects you from the consequences of your own impatience.

And perhaps it's the Vulcans' frontal lobes that saved them from emotional destruction. They may have accomplished what E.O. Wilson was referring to when he said:

> *Although the evidence suggests that the biological nature of humankind launched the evolution of organized aggression and roughly directed its early history across many societies, the eventual outcome of that evolution will be determined by cultural processes brought increasingly under the control of rational thought.*[30]

Wilson was obviously talking about things that happened on Earth, but his words could describe equally well the process that made it possible for the Vulcans to tame their violent tendencies. Let's hope it's not too late for Earth's humans to take a cue from the Vulcans.

What Geordi Saw
and What Quark Heard

The oldies bin at the record store has turned out to be a real gold mine. You've found just the CD you'd been looking for: *Great Tenors from Star Trek*. You've already heard most of the songs, but you can't wait to get the CD home and listen to it. As you expected, the first track is lovely—the rich Irish tenor of Chief Miles O'Brien singing to persuade a renegade Starfleet officer to end his vendetta against the Cardassians (TNG-4, "The Wounded"). The second track is Mr. Spock doing a song from that most musical of *Star Trek* television series episodes, "Plato's Stepchildren" (TOS-3). Then the ridiculous—Lieutenant Commander Data croaking Puccini to his girlfriend (TNG-4, "In Theory")—is followed quickly by the sublime—*Voyager's* holographic doctor displaying his very good tenor voice in a duet from *La Boheme* (VOY-3, "The Swarm"). You wait expectantly as the last track of the CD begins. But you don't hear a thing—just several minutes of silence. You check the liner notes, which assure you that you should be hearing the glorious singing voices of the tenors from the planet Kerelia. Why, then, don't you hear that "glorious singing?" The problem isn't with your CD player—it's your human ears that make you deaf to the Kerelian tenors.

The world of any creature living on any planet is constrained by the limitations of its senses. The particular sense organs possessed by that creature determine its sensory experiences. Our world would be markedly expanded if, for example, we acquired x-ray vision like Superman's—we'd be able to see individual parts within the interior of solid objects. Imagine looking at your hand and being able to see the tiny bones of your fingers! In contrast, our world would be diminished if we lost one of our visual abilities, such as color perception. Without being able to discriminate differing shades of red and green, we'd have difficulty selecting ripe fruits and vegetables at the market.

Our senses are our lifelines to the world—they deliver crucial information about objects and events in our environment. This information may consist of vibrations picked up by the ears ("sounds"), mechanical pressure on the skin ("touch"), light imaged within the eye ("sight"), or chemicals registered in the nose ("odors") and on the tongue ("taste"). Most of the time, what we perceive is the joint product of several senses working together.

Everything we know about the objects and events around us comes courtesy of our senses. However, our sensory organs don't respond to *all* the available information in the environment. They're "tuned" to pick up the information most relevant to our habits and needs.[1] For example, it is unimportant for us to perceive infrared energy, so our bodies have no sensory receptors responsive to this portion of the energy spectrum. But some snakes are exquisitely sensitive to infrared energy, and they depend on this information to locate warm-blooded prey. The design of any animal's senses—ours included—is crafted by natural selection over countless generations.

What does this have to do with those Kerelian tenors you couldn't hear? Kerelian tenors produce musical sounds that fall outside the range of human hearing (TNG-6, "Lessons"). According to Captain Picard, to appreciate the subtle differences in sound that make a Kerelian tenor a virtuoso, you must have extraordinarily sensitive hearing—as all Kerelians do. Human hearing is not up to the task, so we are deaf to their virtuosity.

You don't have to be a Kerelian, or any other kind of humanoid, to have superhuman perceptual abilities. Even Earth's lowly housefly qualifies as superhuman, at least in one respect. When objects in the environment appear or disappear rapidly, *Musca domestica's* eyes respond much more quickly than ours do. Compared with the housefly's, the human eye is glacial in its response speed. Yet what might be construed as a deficiency in human vision in fact has benefits: the sluggish responses of our eyes and brain make it possible for us to appreciate the action in our favorite *Star Trek* film.

Here's how it works. Because your visual system responds relatively slowly when a series of still pictures are projected one after another on a movie screen, your vision fails to keep up with all the changes, blending one image into the next. The result is the illusion of motion and the equally strong illusion that the screen is never blank. Actually, the movie screen is dark between frames. The fly's visual system, being much faster in its responses to change, would have no trouble distinguishing individual frames. To a fly buzzing around the theater, the light reflected from the screen would appear to be flashing on and off with each of the film's million or so frames. The visual effect would resemble the stroboscopic illumination that you may have experienced at a disco.

The Better to See You with, Bones

So, the world viewed through the eyes of a fly looks different from the world we humans experience. But there are also remarkable differences in vision among individual humans. Different humans, in other words, live in different visual worlds. For example, most humans over the age of forty-five experience presbyopia ("old sightedness"), a progressive, age-related difficulty in focusing on near objects. Presbyopia produces some odd symptoms, most notably shortened arms. Even with their arms fully extended, victims of presbyopia can't hold a newspaper or book far enough away for it be readable—the letters on the page appear blurred. Unfortunately, even in the twenty-third century, presbyopia is still a problem, as we learn from a birthday gift to Admiral James T. Kirk, in *Star Trek II: The Wrath of Khan*. At various points in the movie, Kirk's presbyopia makes it difficult for him to carry out his duties. He has trouble reading his wristwatch or the information on the starship's display panels. For a Starfleet officer, this is more than an inconvenience—it's totally unacceptable. Admiral Kirk's visual world changes as he grows older, just as yours will.

To solve Kirk's problem, Dr. Leonard McCoy gives his old friend a pair of spectacles, so-called "half-frame glasses." From the time Benjamin Franklin invented "halfies" in the eighteenth century, glasses like Kirk's have been used to help presbyopic eyes. They work by magnifying the image of whatever the wearer looks at. But why would a *Star Trek* officer centuries after Benjamin Franklin need such a primitive device, particularly when, on a starship, glasses of any sort are as scarce as hens' teeth? McCoy explains that he usually prescribes a drug, Retnax V, to his presbyopic patients. But the drug is counterindicated for Kirk: he's allergic to the stuff.

Retnax V is not listed in the 1998 edition of the *Physician's Desk Reference*. But, from what is known about the origins of presbyopia, we can make some educated guesses about how it must work. Presbyopia results when the eye's lens loses elasticity and, therefore, its ability to change its shape to maintain a focused image on the retina, at the back of the eye. And the nearer an object is, the more focusing must be done, which is why Kirk can see distant objects fine, but not near ones, like his wristwatch. Presumably, then, Retnax V restores the flexibility of the lens, or strengthens the ciliary muscles whose contraction changes the lens's shape to help us focus on close objects, or both. But in our spot on the space-time continuum, we still have to make do with the old-fashioned treatment: a pair of reading glasses.

Just as vision changes with age, so, too, does hearing. The most common type of age-related hearing loss is called presbycusis ("old hearing"). Doing its work gradually over the years, presbycusis diminishes hearing sensitivity, typically beginning at the high-frequency end of the sound spectrum and working its way down the scale as aging progresses. Its first signs are errors in understanding particular sounds in speech, distinctions for which higher-frequency hearing is crucial. The distinction between "sign" and "shine," for example, may be inaudible to someone with presbycusis, but perfectly clear to a younger person whose hearing has not been affected. Perceptual differences of this kind obviously do little to facilitate communication between younger and older generations.

If a twentieth-century human lives long enough, some degree of age-related hearing loss is nearly inevitable. The causes of presbycusis have not yet been worked out, but we do know that some families have a greater genetic predisposition to it. We also know that continuous, lifelong exposure to noise increases the severity of the hearing loss. If the current thinking is correct, the increasing ambient noise of modern society—from automobiles, subway trains, airplanes, and the like—will make presbycusis an even more serious problem in the future. One obvious offender is very loud rock music, like the intense noise emanating from a leather-clad bus rider's boom box that annoys Mr. Spock in *Star Trek IV: The Voyage Home*. For the sake of our own hearing, it's too bad we have no society-wide equivalent of the deftly delivered Vulcan nerve pinch that brings golden silence to the bus (and temporary unconsciousness to the boom box owner).

We have no way of knowing whether Earth will quiet down over the next few centuries, but we do know that starships can get pretty noisy at times. Red alert sirens, phaser discharges, and warp engine explosions are just a few of the sudden loud sounds experienced by Starfleet officers and crew.[2] Studies on mammalian ears, including those of humans, show that sudden loud noises can permanently damage the tiny, delicate hair cells of the ear's innermost chamber. It is these cells that convert acoustic energy into neural impulses that are transmitted to the brain, so they are crucial to normal hearing. Unfortunately, these delicate hair cells can be destroyed by exposure to loud noise; the consequence is permanent hearing loss. Many cases of traumatic hearing loss induced by sudden loud sounds are reported each year in the United States, particularly around the Fourth of July, when celebratory firecrackers take a toll. Perhaps the best-known case of noise-induced hearing loss is that of former actor and U.S. President Ronald Reagan.[3] According to his own account, he started to lose his hearing in 1930, long before he became President, when a gun was fired near his ear while he was making a movie. His hearing never recovered.

So, every time a starship goes into battle, the ears of officers and crew alike are at risk for trauma from the loud sounds bombarding their ears. Unless twenty-fourth-century medicine has devised a way to protect the inner ear against noise-induced trauma, it is nothing but miraculous that no Starfleet crew member shows signs of hearing loss.

It's the Lobes

You cannot discuss hearing and *Star Trek* without thinking of the Ferengi, a species renowned for its acquisitiveness and its prodigious ears. The visible external portions of male Ferengi ears are enormous, and in proportion to their owners' short stature, they seem even more exaggerated.

Ferengi ears have about the same length-to-width proportions as human ears, but in the typical male Ferengi, the linear dimensions of the ear are about three times greater. This makes the *area* of the Ferengi ear nine times that of its human counterpart. And when it comes to catching sound, it's area that counts. Although these huge external ears—"lobes," as they're proudly called by Ferengi— are key erogenous zones for male Ferengi (TNG-3, "Ménage à Troi"), they are also interesting as organs of hearing.

Star Trek's most familiar Ferengi is Quark, the resourceful rascal who runs the bar on the space station Deep Space Nine. And a lot of what we know about Ferengi hearing comes from Quark. In "Looking for par'Mach in all the Wrong Places" (DS9-5), Quark discovers Dr. Julian Bashir trying to eavesdrop on an argument between Miles O'Brien and Major Kira Nerys. Unfortunately, the door to the room where the argument is taking place muffles their words to the point of inaudibility. After all, Bashir is human, and his human ears are simply not up to the job. Quark is quick to lend a lobe. His hearing is so sensitive that he doesn't even have to put his ear to the door. Standing several feet away from it, Quark is able to hear everything. Although Ferengi are not generally known for their generosity, Quark shares with Bashir every juicy detail he has overheard, proud of what his prodigious lobes can accomplish with such ease.

There is no question that big ears like Quark's do improve their owner's hearing. To see how that improvement comes about, consider the human ear. The external ear, technically called the auricle, acts like a funnel, capturing airborne acoustic energy and channeling it into the external auditory canal (the finger-sized opening within the shell-shaped auricle). To hear very weak sounds (like the muffled voices of two people arguing), the ear has to catch as much of the airborne energy as possible. The more energy captured, the easier it is to hear. In a way, Ferengi lobes resemble the very first low-tech hearing aid: a hand cupped behind the ear. Though this maneuver enlarges the ear's catchment area for sound, a cupped hand still leaves the human ear far below male Ferengi proportions. Because of their superior sound-capturing lobes, male Ferengi can hear weak sounds inaudible to humans—courtesy of their superior hearing, they live in a richer, more complex auditory world. But their superior hearing carries a price: noises that don't bother humans at all can be unbearably loud to a Ferengi, as we witness when Quark writhes in pain when exposed to a sonic probe that has no effect on Chief O'Brien's ears (DS9-2, "Playing God").

With ears, however, as with other body parts, size is not everything. Although size makes Ferengi ears more receptive to sound, the way they protrude from their owners' heads may be just as important to the exceptional sensitivity of Ferengi hearing. Human ears lie flat against the head, but Ferengi ears stick out, creating a pronounced cup. This arrangement amplifies the directional selectivity of a Ferengi's hearing. To see this for yourself, face some source of continuous sound (for example, the humming hard disk drive of a personal computer), place your hands behind your ears, and push their outer edges forward. Notice the increase in the sound's loudness. Ferengi wouldn't have to push their ears forward—the natural shape of their ears already does this for them. It improves their hearing by capturing sound in greater abundance; in contrast, sounds that the ears are turned away from tend to be screened out. This supercharged hearing comes in handy when one Ferengi serves temporarily as communications officer on board the U.S.S. *Defiant*. Despite the intense noise of battle, this Ferengi has little trouble hearing Captain Sisko's orders and communicating them to others (DS9-5, "For the Uniform").

Or Is It?

Ferengi males take justifiable pride in the size of their lobes, but at least one Ferengi overestimates his lobes' actual contribution to his hearing. In "Darkness and the Light" (DS9-5), Major Kira receives a number of threatening audio transmissions related to a series of murders. Because the noisy sounds of the messages are garbled, neither Kira nor Jadzia Dax, the station's science officer, can identify the speaker or understand the message.

In desperation, Kira and Dax turn to Nog, Quark's nephew and the first Ferengi cadet ever to attend Starfleet Academy. Nog listens to taped replays of the scrambled messages and quickly realizes that they are composites spliced together from different originals—he can hear that the intonation and phrasing are abnormal. Once his superior hearing uncovers the altered nature of the messages, Nog identifies the speaker as a Bajoran female, and this information allows Dax to unscramble the messages and determine their source. Nog attributes his feat of auditory acuity to his oversized Ferengi ears, remarking casually, "It's the lobes." But if human hearing is our guide, Nog is probably referring to the *wrong* lobes.

Recognizing an unseen speaker involves more than raw ability to detect sounds. It requires keen auditory discrimination—the ability to make subtle distinctions among sounds, recognizing telltale clues such as the speaker's accent and rhythm. During the early stages of presbycusis, older people can have fairly normal sensitivity (meaning that they hear a sufficient range of sounds), but substandard discrimination (they confuse speech sounds).

They're able to hear sounds, including voices, but those sounds may be garbled to the point of unintelligibility.

Even people with completely normal hearing find themselves in situations in which messages are heard but not understood. Imagine yourself as a tourist standing on the platform of a subway station in New York City. Suddenly an announcement comes over the loudspeaker—aptly nicknamed a "squawk box." Although you hear the sound of the announcement, other sounds—noise from the approaching train, ongoing loud conversations, subway musicians—conspire to render it incomprehensible. Worried that you've missed some essential information, you turn to a stranger and ask, "What'd she say?" Looking up from his newspaper, the stranger tells you, "The Q Express to Coney Island is being rerouted to run as a local, on the tracks normally used by the D train." Your jaw drops in amazement. How did he make out all of that from a bunch of meaningless squawks?

The stranger's ears are the same size as yours, so his lobes are not the answer. The key? He has had a lot more experience than you have listening to subway announcements under these noisy conditions. In addition, the stranger probably has more knowledge than you do of the kinds of messages that are likely to be heard on the subway platform. Experience and prior knowledge enhance the performance of all our senses, hearing included. But experience, while capable of improving our hearing discrimination, doesn't alter our ears. The lobes that experience alters are the lobes of the brain.

The Brain Has Lobes of Its Own

The outer portion of the human brain—the cerebrum—is organized into four sections called lobes (shown in the illustration on the following page). These lobes—occipital, temporal, parietal, and frontal—are named for the adjacent bony plates in the skull that cover them. The four of them come in pairs, one member in the left hemisphere and the other in the right.

Groups of neurons in each lobe have specific responsibilities. For example, neurons in portions of the temporal lobe receive and process signals originating in the ear. As a result, these temporal lobe neurons are crucial to hearing. Within this portion of the brain, different neurons respond preferentially to different sound frequencies. So as you listen to *Star Trek: Deep Space Nine's* theme

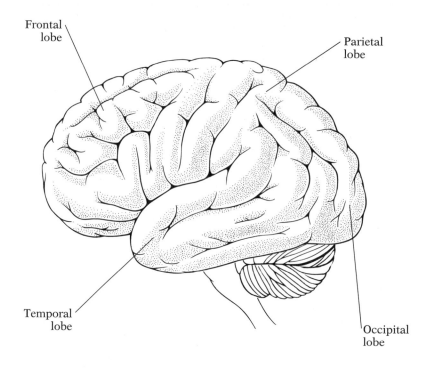

Frontal lobe

Parietal lobe

Temporal lobe

Occipital lobe

music, the various frequencies in the melody activate different temporal lobe neurons, giving rise to the musical sounds you experience.

Recently, neuroscientists have discovered that the brain's responses to sounds are not fixed and inflexible. Instead, they turn out to be easily and quickly changed by experience. One study of this phenomenon used monkeys, whose brains are highly similar to the human brain.[4] The monkeys were trained to discriminate between two similar tones—for example, the tones you hear if you strike the high B and the high C on a piano keyboard. The monkeys could press one of two levers to indicate whether two successive tones sounded the same or different in pitch. Correct responses were rewarded with a sip of orange juice. Despite this powerful incentive, the monkeys initially found this discrimination task difficult, and made many errors. They improved with practice, however, discriminating finer and finer tone differences. Interestingly, their improvement was restricted to tones similar to the ones they experienced during training; for example, training with high-frequency tones had no effect on the monkeys' subsequent ability to make subtle distinctions between low-frequency tones.

While these monkeys were receiving training on the tone discrimination task, neuroscientists were monitoring the activity within individual brain cells in the animals' temporal lobes. Over the course of the training period, more and more neurons became responsive to the tones being experienced by the monkeys. Their brains, in other words, allocated additional neural processing resources to those sounds that were particularly important to the animals, at the sacrifice of neurons responsive to infrequent or unimportant sounds.

This capacity for adaptive modification, known as *plasticity*, is not limited to the neural centers for hearing. Neural plasticity is a built-in property of most areas of the brain, and it allows us to adapt to the demands of our environment. As you learn to type, the portions of your brain responsible for controlling your fingers' movements are reprogrammed. As you acquire the ability to discriminate California Chardonnay from French Chablis, the sensory portions of your brain devoted to taste and smell are reprogrammed. Neural plasticity also allows some degree of recovery from damage to portions of the brain. Sometimes stroke patients can recover lost functions through rehabilitation training—these patients are "teaching" functioning parts of their brains to take over the duties previously handled by the now damaged areas. In effect, the brain behaves like a flexible computer whose circuits can be modified to adapt its owner to current environmental demands.

Eavesdropping Practice

Armed with this information about brain plasticity, let's return to Cadet Nog. Is there reason to think that the lobes in his brain are responsible for his unusually keen hearing? Is there something in his experience that would favor extraordinary perception of speech sounds?

We know that Ferengi lives are governed by the cultural wisdom that has been distilled down to the several hundred Ferengi Rules of Acquisition. (The 6th Rule is a typical nugget of wisdom: "Never allow family to stand in the way of opportunity.") To get the financial edge mandated by the Rules of Acquisition, Ferengi live by their wits and cunning. They are rewarded for taking in all that goes on around them, including conversations that are not intended for their lobes. Quark is a good example. Tending the bar on Deep Space Nine allows him to monitor conversations and gather data that could increase his personal net worth. Synthehol, the most common beverage served at the bar, loosens inhibitions—and also tongues. In "Transfigurations" (TNG-3), Lieutenant Worf complains about this effect, exhorting his companions, "Less talk, more synthehol. We came here to relax."

Thanks to the synthehol Quark serves his patrons, he is able to overhear all sorts of valuable tidbits, such as which commodities are scarce in various sectors of the galaxy. To benefit from his position, Quark must understand his patrons' private conversations, no matter how noisy the bar is. Certainly, the directional selectivity of his protruding Ferengi ears screens out some of the background noise that would otherwise interfere with his eavesdropping (assuming that he tilts his lobes appropriately). But equally important are his years of practice at eavesdropping. Like the stranger in the subway station, Quark has learned to discriminate highly similar sounds that others might confuse.

But what about Nog, the young Ferengi whose keen hearing decodes those threatening messages? As a boy working in his uncle's bar before he goes off to Starfleet Academy, Nog shares his species' keen interest in what other humanoids are talking about. With no disrespect meant to Ferengi, Nog can be likened to the trained monkeys whose brains were altered by practice at listening to tones. Intense practice at his own listening task—listening to conversations that he isn't supposed to hear—has probably promoted reorganization of his brain's temporal lobes, encouraging unusually large numbers of nerve cells to respond to the kinds of sounds that make up humanoid speech.

A Touching Experience

So far, we've focused on adaptive brain processes in monkeys and Ferengi. What can we say about brain and sensory plasticity in our own species? Some of the most intriguing information about plasticity in human brains comes from studies of the sense of touch. Unlike the other senses, whose receptors are confined to a small part of the body (such as the back of the eye or the surface of the tongue), the receptors for touch are scattered over the entire surface of the body. Our skin is dotted with thousands of specialized receptors that generate electrical responses whenever the skin contacts an object. These so-called mechanoreceptors are especially numerous in the skin covering the fingertips and the lips, two regions with exquisite sensitivity to touch.

Signals generated in the skin's mechanoreceptors travel over nerve fibers to the brain, ultimately arriving within an area of the parietal lobe known as the somatosensory cortex. (The *somato-* part of *somatosensory* means "body.") This region is located roughly halfway between the brain's front and back, slightly forward of and about six centimeters above the middle of the ear. Each individual neuron in the somatosensory cortex responds to touch stimulation at a particular location on the body. Touching the left thumb, for example, activates different neurons than does touching the left index finger. It is worth noting that neurons in the somatosensory cortex of the left hemisphere monitor touch stimulation of the right half of the body, and neurons in the right hemisphere register touch sensations from the left half of the body. As a result of this criss-cross arrangement, when you touch something with the fingers of your right hand, information goes from your fingertips up the nerves in your right arm, into the spinal cord, and ultimately to the somatosensory region on the left side of your brain.

Together, the millions of somatosensory neurons constitute a "map" of the body surface. Within that map, different parts of the body command different numbers of neurons. This map can be put into pictorial form by drawing the size of a body region in proportion to the size of the brain region devoted to its representation. Because the resulting picture looks something like a human form, this representation is called a homunculus, which means "miniature man." In the homunculus associated with the human sense of touch, the tongue, hands, and genitals are all outsized, reflecting the high touch acuity associated with these body regions.

The somatosensory portion of the brain—like the brain's auditory areas—can be modified by experience. Consequently, people who use touch in radically different ways will have brains whose somatosensory neurons are organized differently. Take, for example, professional musicians who play stringed instruments. These individuals rely much more heavily on the left hand—which must depress the strings in precise locations—than they do on the right hand—which merely bows or strums. In terms of how they use the fingers of their two hands, string players are not like the rest of us, particularly those of us who are right-handed. But what about the brains of these musicians? Are they different as well?

As described in Chapter 2, modern brain imaging techniques can provide a picture of those brain regions that are activated in response to specific events. Suppose, then, that we measure the extent of brain activity in response to tactile stimulation of the fingers of the left and right hands. In most people, the left and right hands enjoy comparably sized representations within the somatosensory regions of the right and left hemispheres, respectively. But in musicians whose instruments place greater demands on the left hand, the touch representation of the left hand is much greater than that of the right hand.[5] In other words, the brain's touch map has changed in response to the musician's tactile experiences. And musicians who began their training early in life have larger left-hand brain maps than do those who began their training later. Evidently, early onset of string playing recruits more brain cells to serve the touch sensors in the critical left-hand fingers. So if you aspire to be a virtuoso on the violin, or the electric guitar for that matter, don't waste any time. Get going immediately, or at least as soon as you finish this book.

What would the homunculus for a Ferengi brain look like? Because a male Ferengi's lobes are key erogenous zones, it's a safe bet that touching them would elicit strong activity over a very large expanse of the somatosensory region of his brain. Certainly the Ferengi homunculus would have gargantuan ears. We can only speculate about its other attributes, but the principle certainly would apply: Body parts that are important to their owner, or body parts that enjoy extra use, recruit extra neurons in the brain.

Two Blind People with Terrific Vision

A *prosthesis* is a device that compensates for a missing or injured portion of the body. False teeth qualify as a prosthesis, as do the reading glasses Kirk gets for his birthday. *Star Trek* has introduced us to protheses that augment physical strength, such as Ensign Melora Pazlar's servo-harness (DS9-2, "Melora"). This device helps Pazlar move about in gravitational forces that are normal for *Homo sapiens*, but much stronger than those on her home planet. A related prosthesis, the motor assist device, enables Worf to walk after his spinal cord is damaged. A broken spinal cord shuts off communication between the brain and limb muscles; motor assist devices on the limbs pick up the brain's command signals and send them to the appropriate muscles (TNG-5, "Ethics").

Star Trek prostheses are not just about muscle power, however. Sometimes they bestow superior mental ability. For example individual Borg, humanoid-machine hybrids, exploit various prostheses, such as sinister-looking electronic eyes. As a collective, the Borg share a prosthesis, a communications network that allows each Borg to share the knowledge and information of the others.

Certainly among the most interesting of *Star Trek's* prostheses is the one that gives sight to blind psychologist Miranda Jones (TOS-3, "In Truth Is There No Beauty?").[6] When Dr. Jones arrives on the *Enterprise*, she is accompanied by the Medusan ambassador Kollos. Like other Medusans, Kollos is so ugly that the sight of him can drive humans mad.[7] Dr. Jones's blindness protects her against this fate.

But when Dr. Jones beams aboard the *Enterprise*, she shows no signs of blindness. She moves about gracefully in unfamiliar surroundings, without hesitation or error. She is not slowed down at the dining table, and she maintains eye contact with people to whom she speaks. She even spots the tiny IDIC symbol on Spock's jacket.[8] Dr. McCoy has advance information that the visitor is blind, but the rest of the crew hasn't a clue. How does a totally blind person carry off this remarkable deception? She has a very good dressmaker.

Dr. Jones's entire dress is covered by a large prosthetic device: a flexible net dotted with hundreds of large, shiny beads. While they appear to be decorative ornaments, these beads are actually sophisticated optical sensors that act as her eyes. Without the web of beads, she is literally and figuratively in the dark. We never learn how Dr. Jones's beads work, but clearly they do.

The simplicity of Dr. Jones's prosthesis presents several thorny technological challenges. For one thing, the absence of sensors above her shoulders makes it hard to understand how she gets visual information about events near her eye level. Yet that is exactly the information she needs for the excellent eye contact she maintains with Kirk, Spock, and the others. Also, the sensor beads are attached to a form-fitting mesh web that Dr. Jones puts over whatever dress she wears. This means that each time she dons the sensor web, the position and orientation of each sensor changes. This creates a challenging problem. The sensor web can be compared to a large array of security cameras scattered throughout the hallways of a large building, which display their images on a bank of television monitors located in a central security office. If the security officers don't know the location of the camera represented by each monitor, they can't interpret the information they receive. Similarly, to make sense of the information it gets from the sensor web, Dr. Jones's brain would have to know, moment by moment, the precise position of each sensor. As a result, her web sensors would have to be recalibrated every time she gets dressed—and whenever she sits down or stands up.

Dr. Jones's substitute "eyes" are so subtle and unobtrusive that the *Enterprise's* crew does not catch onto her blindness. In contrast, *Star Trek's* most famous visual prosthesis, Commander Geordi LaForge's VISOR, is impossible to miss. The VISOR (Visual Instrument and Sensory Organ Replacement) is a shiny metallic band that covers LaForge's eyes. Blind from birth, LaForge has had a visual prosthesis since he was five years old (TNG-5, "Hero Worship").

The VISOR allows LaForge to see, but his visual experiences do not resemble those of sighted people. By routing signals from the VISOR to a video screen, the *Enterprise*-D's other crew members are able to see a simulation of what he sees (TNG-1, "Heart of Glory"). This simulation dramatizes how different his visual world is from theirs. When LaForge fixes his gaze on Commander Riker, Picard sees Riker's transmitted image as a vague, ill-defined shape amid patches of shifting colors. Picard struggles for words to describe this novel sight, finally settling for, "To me it's just a . . . an undefined form standing in . . . in a visual frenzy."

LaForge shrugs off the difficulty of interpreting this visual image, saying that it's just a matter of practice, that he selects what he wants to see and disregards the rest. His unique, VISOR-eye view of the world reminds us that the way the world looks depends upon the eyes and the brain that are "seeing" it. For the crew, and for us, the chance to see through LaForge's eyes is a powerful reminder that the *physical* world—the stuff we can measure with physical instruments—is not the same as the *perceptual* world—the stuff we experience.

Simple, Appealing, and Flat-out Wrong

People often have difficulty appreciating the difference between the perceptual world and the physical world. It is easy to believe that the two are the same. This view is so common that it has a name: *naive realism*. A card-carrying naive realist believes that his perceptual experiences accurately and completely define the objects and events in the physical world. To the naive realist, a rainbow *appears* colored because it *is* colored; a metal cup *feels* cool to the touch because it *is* cool.

This view of perception is simple, appealing—and flat-out wrong. Several lines of evidence disconfirm naive realism. For one thing, perception can fluctuate over time, even though the physical stimulus being perceived remains unchanged. The illustration shown below exemplifies the unstable nature of perception. Look for a few seconds at the lone cube in the insert at the upper right. At one moment, its line segment AB seems to be closest to you, while at another moment, segment CD seems closest. If you look at the cube in the center of the nonet, you will probably see that this fluctuation is not an isolated phenomenon: when one cube fluctuates, its fellows go along. There would be no such shape-shifting at all if perception were determined *only* by the physical properties of what we look at. But our perceptual experiences include qualities that are not present in the physical world.

Naive realism is also rebutted by the fact that perception varies with context. How an object appears depends not only on

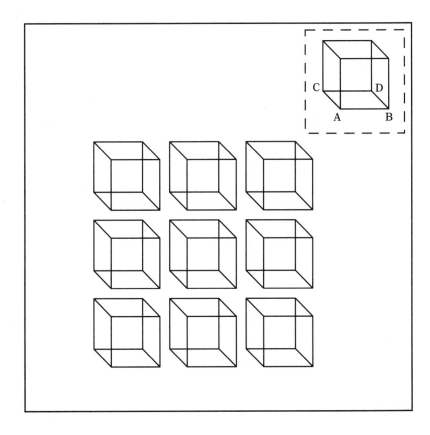

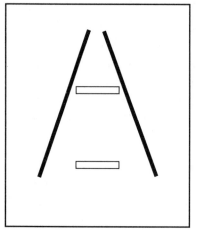 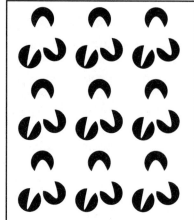

its physical characteristics, but also on the characteristics of whatever other objects may be in its vicinity. Look at the pair of horizontal lines in the box at the left of the above illustration. Which is longer, the upper one or the lower one? Most people would say the upper one, yet both are identical in length. The pair of long, dark, slanted lines creates a misleading context that diverts perception from the straight and narrow path.

When you look at the right-hand box in the illustration, you probably see a series of white triangular objects. They probably resemble the triangular emblems on Starfleet communicator badges. But the physical reality is different. A photometer, a device that measures the physical distribution of light, would reveal no evidence that these emblems exist on the page. They are subjective, illusory contours, fabricated by neurons in your brain's visual areas. But don't be alarmed. Your brain is not being perverse—at least no more so than usual. This process of fabrication allows you to see complete, whole objects even when some of their parts are hidden from view. Make a cross with your two index fingers, placing one on top of the other. You see two complete fingers, even though a portion of one finger is obscured from sight. The illusory contours fabricated by your brain arise from the same neural operations that deduce the existence of a complete finger. The brain may sometimes lie, but it's an intelligent liar.

Here's another example of the brain's propensity to misrepresent matters for good reason. The eyes of a typical human blink shut, on average, about four or five times per minute—more if the person is tense or is telling a lie. Each blink closes the lids for about a quarter of a second. By nervous system standards, a quarter of second is a long time. If a computer or television screen blinked off for that long, we would see it very easily. But we never notice the effects of our constant blinking. Because it's responsible for producing those blinks, the human brain knows it can ignore transient "off" periods when they are self-produced—we never notice that our blinks are constantly turning out the lights. And this is fortunate, so that we don't confuse self-produced blackouts with real ones.

These various optical illusions all undermine the plausibility of naive realism. Geordi LaForge presents another strong, concrete argument against naive realism: looking at the world through his VISOR, we realize that when different people look at the same visual scene, not everyone necessarily sees it the same way. We know, for example, that young human infants cannot see small objects that are clearly visible to adults. An infant's visual nervous system has not matured sufficiently to register fine visual detail. We also know that people with deficient color vision confuse hues—such as reds and oranges—that others, with "normal" color vision, distinguish with no effort at all.

Analogous individual variations exist in the other senses as well. To give an example, disease, genetic accident, or head trauma can render some people completely insensitive to odors that generate strong olfactory experiences in the rest of us. These facts challenge naive realism—despite living in the same physical world, different creatures live in different perceptual worlds. LaForge's VISOR underscores this disparity between the physical and the perceptual. It gives us a unique opportunity to see the world through someone else's eyes. But what does that world look like? Before answering that question, we must know a little more about the VISOR itself.

How Does the VISOR Work?

To make sight possible, the VISOR would have to register the distribution of electromagnetic radiation within its field of view. Based on what is known about the eye and the visual portions of human brain, we can make some educated guesses about h' device works. The VISOR probably contains an array of ' sensors, each registering the level of electromagne' arriving from a particular portion of space; togetl.

sensors defines the VISOR's visual field. Electrical signals generated within the sensors are sent to a pair of electrical terminals, one slightly behind each eye's outer margin. From these so-called neural nodes, the signals are forwarded to the nethermost region of LaForge's brain: the visual cortex within the occipital lobe.[9]

To understand the next stage in the process, you need to know something about what happens when neurons are activated in a human's visual cortex. Electrical stimulation of neurons in the human visual cortex evokes clear sensations of light—visual impressions known as phosphenes.[10] One famous case involved an adult woman who had lost the use of her eyes through disease. Her physicians wanted to learn whether visual sensations could be evoked by direct stimulation of her brain, which remained unaffected by the disease. They implanted an array of tiny electrodes in the visual cortex of the patient's brain. When minute electrical currents were sent through the electrodes, the patient reported visual experiences. Depending on which neurons were activated, she described seeing small spots of light resembling "stars in the sky" or short lines that looked like "half a matchstick at arm's length." Similar studies have since been performed on other humans, and their results point to the same conclusion: stimulation of different visual neurons elicits different visual sensations. Specific colors are associated with the stimulation of some neurons, while others generate impressions of white figures. A given neuron reliably elicits the same visual impression, and each neuron's visual image appears within a particular location in the visual field.[11] Moreover, stimulation of neurons located near one another evokes visual impressions in neighboring regions of the visual field. These effects are predictable enough that the simultaneous activation of several different sets of neurons is able to produce the sensation of a more complex figure, such as a letter of the alphabet.

Medical scientists have succeeded in exploiting phosphenes to bring crude vision to a few blind human volunteers. An optical device resembling a small video camera converts patterns of light into patterns of electrical signals in an array of electrodes implanted in the volunteer's brain. With this arrangement, a vertically oriented bar of light appearing within the middle of the camera's field of view would activate a tiny cluster of neurons in contact with the electrodes associated with that region of the camera; a horizontal bar of light located slightly higher in the visual field would activate a neighboring cluster of neurons. Together, these patterns of activity might create the impression of the letter *T*. Indeed, in some experiments, this device made it possible for blind human volunteers to read. It should be stressed that these technological developments are still very much in their infancy. But assuming that *Star Trek* provides a glimpse into the future, such devices have great promise.

LaForge, Master of the Invisible

The information delivered to LaForge's brain is very different from what our brains receive from our eyes.[12] To understand the uniqueness of his vision, we must turn to the physics of light. Whenever an electrical charge oscillates or accelerates, it sets up a disturbance called electromagnetic radiation (EMR). This disturbance spreads out (radiates) in all directions, creating an electromagnetic wave. Light is one particular kind of EMR; x-rays, radio waves, and microwaves are others. EMR is categorized according to its frequency, or according to the reciprocal, its wavelength (in fractions or multiples of meters).

Normal human vision depends on a tiny band of the huge EMR spectrum. This sliver of radiation, which we call light, includes only wavelengths between 400 and 700 nanometers. Ordinarily, humans see objects in the environment because light from those objects has reached the photoreceptor cells of the retina. The light picked up by these receptors has either been emitted by the objects or, in the case of nonluminous objects, generated elsewhere and reflected by them. A firefly's glow is one example of emitted radiation, and an image on a computer screen is another; the patterns of light and dark on the pages of this book represent reflected radiation.

When an object does not emit or reflect sufficient energy within the bandwidth 400 to 700 nanometers to stimulate human photoreceptor cells, that object is invisible to the human eye. This simple truth is captured in Leibowitz's Law (whimsically promulgated by vision scientist Herschel W. Leibowitz): "One cannot see a damn thing in the dark." Commander LaForge is the only human for whom Leibowitz's Law is inoperative: he can see perfectly well without light, as long as there's other EMR.

Our photoreceptors' limited sensitivity also makes us blind to radiation other than light. For example, we cannot see ultraviolet radiation—EMR whose wavelengths range from 4 to 400 nanometers—or infrared radiation—EMR in the bandwidth 700 to

1,000,000 nanometers. However, twenty-fourth-century technology allows LaForge to overcome these normal human barriers to sight. He says this about his VISOR's bandwidth:

> *It's a remarkable piece of bioelectronic engineering by which I, quote, "see" much of the EM spectrum, ranging from simple heat and infrared through radio waves . . .*

> (TNG-1, "Encounter at Farpoint," II)

He is absolutely right in putting quotation marks around the word "see," because what he experiences is not what normally sighted humans mean by "seeing." His VISOR is sensitive to a far greater range of radiation than our eyes are, enabling him to "see" not only light, but also energy from other regions of the EMR spectrum. When LaForge looks at Commander Riker, the VISOR registers not only the light reflected from Riker's body, but also the infrared radiation (heat) that Riker emits. The VISOR's sensitivity to various forms of radiation causes double or even triple images to be sent to LaForge's brain, each image representing a different part of the EMR spectrum.

On occasion, the VISOR's ability to register diverse forms of EMR is extremely useful. Consider what happens when LaForge is part of an away team that beams to a severely damaged Talarian freighter (TNG-1, "Heart of Glory"). Inspecting the freighter's hull, he sees a stress pattern in the metal that warns him that the hull is losing its structural integrity and is about to crack. This stress pattern is invisible in light, but shows up in another part of the EMR spectrum. Because the VISOR allows LaForge to see the stress pattern, the team is able to escape before the hull blows. The VISOR comes in handy again in "The Enemy" (TNG-3), when its sensitivity to high-energy particles allows LaForge and a companion to follow a homing beacon to safety, despite conditions of zero visibility. The VISOR's extraordinary sensitivity also serves less serious purposes, as when LaForge is able to "read" the special infrared-transparent playing cards of his fellow poker players (TNG-5, "Ethics").

Under very special conditions, humans can have their vision extended into a region of the EMR spectrum that they normally do not see. Usually, the lens of a human eye screens out a good deal of the ultraviolet radiation that strikes the eye. This screening keeps this potentially damaging radiation from getting further inside the eye, where it might do harm to the delicate photoreceptors. If the lens of the eye is removed, however, ultraviolet radiation can reach the retina, where, in sufficient amounts, it can evoke a response from some photoreceptors. This is exactly what happened when a cataract (a lens opacity) was removed from the eye of a very observant physician, Robert Anderson, who wrote vividly about his experience.[13] Removing the cataractous lens from the eye also removed the lens's pigment, which normally screens out most of the incoming ultraviolet radiation. With lens and pigment out of the way, ultraviolet radiation was able to reach Dr. Anderson's retina. What did he see?

Using the eye from which the cataractous lens had been removed, Anderson could see not only the light, but also some of the ultraviolet radiation reflected by objects in the environment. The result took some getting used to. For example, his wife bought him some socks that had an indescribable bright color when viewed with his operated eye. The same socks were a muted brown when viewed with his normal eye. This small incursion into a part of the EMR spectrum normally hidden from humans did not elevate Anderson to Geordi LaForge's level of EMR diversity, but his discovery about his own sight should remind us that there is potentially more to the visual world than normally meets the eye.

LaForge "Sees" Better Than He Should

Judging by the video display from the VISOR, Geordi LaForge's prosthetic device gives him a fairly normal-sized visual field (TNG-1, "Heart of Glory"). This means that at any single instant, with his gaze fixed, he sees a chunk of the world that is about as wide and as high as what normally sighted humans see. But the video display also reveals that LaForge's visual field differs from ours in its uniformity of detail. The display from the VISOR has just as much detail near its edges as it does at its center. In contrast, normal human vision delivers many times more detail from the center of the field of view than from its edges.

Perception of visual detail is commonly assessed in terms of acuity, a measure of how sharply the eye sees. The familiar term "20/20 vision" refers to one common index of visual acuity. Usually, visual acuity is assessed by reading letters on an eye chart—the smaller the letters you can read at a certain distance from the chart, the better your acuity.

Visual acuity varies dramatically from one place on the retina to another. As you read these words, the letters you are looking at are imaged on the center of the retina, within the fovea, a highly specialized region in which a large number of photoreceptors are densely packed together. The fovea affords the eye's best visual acuity. As you read, your eyes move, keeping the fovea aimed directly at the words of current interest.

When researchers measure an eye's visual acuity, they find that it worsens as the region tested shifts away from the fovea to more peripheral areas of the retina. To be recognized in our peripheral vision, a letter of the alphabet has to be several times larger than necessary for central vision. If the letter is not large enough, it will be indistinct and blurry. You can experience this for yourself. If you look at the asterisk in the center of the box shown below, you'll see that all the letters and digits around it look equally sharp. But are they really? While keeping your eyes rigidly fixed on the asterisk, try to read the letters and digits two lines down from it, or half a dozen characters to the right or left of it. Remember: Keep your eyes still as you check this out.

The box makes a point that may be hard to accept. It reminds us that what we experience is not a faithful, straightforward translation of the information coming from our sense organs. The apparent uniform sharpness of our visual world is an illusion, a fabrication by the brain. When we look straight ahead at one particular object, it appears crisp and well defined. But so do

5lm484dk5ha63tes3ea04fcm8v1
2nq2hhs48aidiciug93hd7hh38bf72
37gu84ng0e84nc957gn40fjv8n6
1h374hc08h73*e5ndfg8h74h5fd
0jnS8fn30as81h203akoqw893w
19bfs78cw92nc139gn45tng8jlp1
g983h49hxdfkypq8w37885af33idic02

all the other, surrounding objects, even objects seen in our peripheral vision. Yet we have learned from the reading test that objects are nowhere near as distinct in our peripheral vision as they are in our central vision. Letters and digits away from the center of our gaze were hard to read. So here is a paradoxical state of affairs: The letters on a page do not vary in sharpness, but the eye's neural encoding of those letters does. Despite the variation in sharpness inside our eyes, we experience uniform sharpness.

This reading test reveals that one region of the retina is specialized for very good acuity, with the rest of the retina lagging behind. What is the rationale behind this division of labor? Probably anatomical economics and a matter of priorities. Each human eye contains about 125,000,000 photoreceptor cells. Signals from each eye are carried to the brain by about 1,250,000 fibers in the optic nerve. Because photoreceptors outnumber nerve fibers by about a hundred to one, the retina has to compress the information generated by the photoreceptors to accommodate the limited number of fibers in the optic nerve. Users of personal computers will probably recognize such neural recoding as akin to the data compression that is often used to reduce file size for more efficient storage on a disk or faster transmission over a network. If the retina did not compress its data, the flood of visual information from the photoreceptors would require an optic nerve containing 100 times more fibers than it now contains. The optic nerve would have to be so thick that the skull would have to be significantly redesigned to accommodate it. The retina's data compression avoids this problem. It is so efficient that little significant information is lost, and the human brain is quite adept at making optimal use of the information it does receive—or, as in the case of the subjective contours above, extrapolating from that information.

In engineering terms, the VISOR does relatively little preprocessing of the signals it delivers to the brain.[14] "Preprocessing" collates raw, undigested data—including neural signals—into rough, but useful, categories. In the retina, preprocessing groups together similar neural signals, setting the stage for further, more sophisticated analysis. This makes for faster signal transmission and more economical use of neural hardware. Because every known biological sensory organ does considerable preprocessing, it seems odd that the VISOR doesn't follow suit. Certainly, this twenty-fourth-century design decision does not come from technological backwardness. Using our own present-day technology, engineers have produced electronic circuits that can mimic a great deal of the human eye's preprocessing. These circuits are etched on a tiny electronic chip that acts like an artificial human retina.[15] With a television camera acting as the chip's eye, the chip can solve rudimentary visual tasks, such as identifying the borders of objects, or even helping to steer a car on the highway.

Without talking with the VISOR's designers, it is impossible to know why they didn't incorporate more preprocessing, but we can offer an educated guess. Think about what your eyes are doing as you read the successive words in this paragraph. They sweep quickly from left to right, jumping down a line and back to the left when they near the right-hand edge of the page. These rapid gaze shifts keep the images of the letters positioned on your fovea, the central area of the retina where visual acuity is best. We normal humans have to move our eyes because letters are legible to us only when we are looking directly at them. LaForge's VISOR probably has no gaze-shifting circuit, and probably doesn't need one. As we have seen, judging by the video output from the VISOR, LaForge's vision is equally sharp over his entire visual field. Thus, he can do without the frequent gaze shifts that the rest of us depend on. If LaForge read the words on this page, his head and VISOR could remain perfectly still. Of course, when he wants to look at something outside his field of view, he turns his head. Not high-tech, but it works.

Seeing Things My Way

However much his vision differs from everyone else's, LaForge accepts his own way of seeing as the norm.[16] It's what we all do. We automatically assume that everyone else sees the same thing we see. But this assumption isn't always correct. For example, members of some African tribes fail to see visual illusions commonly perceived by Westerners. Remember the line-length illusion we showed you on page 112? When villagers in rural Uganda were shown the illustration, they saw the two horizontal lines as *equivalent* in length—their perception of line length was uninfluenced by the other elements in it.[17] Why are we susceptible to the illusion, but the villagers immune? The answer may have to do with experience. The illustration is, after all, a flat, two-dimensional drawing. Yet we see depth in a drawing when its artist uses perspective cues to portray objects. For example, in the illustration, the two slanted lines can be interpreted as two parallel lines receding in depth. The two horizontal lines are then seen at different depths themselves, leading us to believe that they're different in size. Through a lifetime of exposure to such cues, we learn to perceive depth where, in fact, there is none. Perhaps Ugandan villagers have less experience viewing drawings than we do, so the illustration's perspective cues didn't fool them. Evidently, experience plays a role in shaping what we see, just as it does in the cases of hearing and touch.

Still, it is startling when someone else insists that he sees things differently from you. In "Heart of Glory" (TNG-1), Picard is mesmerized by the video display of what LaForge is seeing. Hoping to understand what LaForge sees, Picard instructs him to shift his gaze to the *Enterprise*-D's second officer, Data. Data is an android, made of electronics and synthetic materials rather than flesh, bone, and blood.[18] His outer covering does not absorb and reflect EMR as human skin does. This difference is invisible to normal human vision, but not to LaForge's VISOR-mediated sight.

> PICARD: Look over at Data. [Picard sees a blurred, indistinct, dark form on the video display. The form has a lighter halo around it; he is surprised.] There's an aura around him.
>
> LAFORGE: [Laughs] Well of course: He's an android.
>
> PICARD: You say that as if it's what we all see.
>
> LAFORGE: Don't you?

Even LaForge treats his own sensory experiences as the gold standard, and is surprised to discover that not everyone sees things his way. Apparently, even in the twenty-fourth century, people—including some very smart people—find it hard to put themselves in another person's shoes—or behind another person's eyes!

6

Even Starship Captains Get Amnesia

It's a medical nightmare. The patient, in critical condition, has been prepped for a delicate operation to repair damage to the motor cortex of his brain. The highly skilled doctor is confident, scrubbed up, and ready to go. Then, just as he's about to begin the operation, the doctor lowers his instrument. He has completely forgotten how to carry out the operation. Embarrassed, he tries to cover up his memory failure by pretending to test his nurse's knowledge of the operation. Recognizing what's wrong, the nurse—her name is Kes—steps in to save the day, and the patient. Step by step, she guides the doctor through the procedure. Over the next few days, the doctor's memory deteriorates steadily, failing in one area after another. Despite occasional moments of temporary recovery, the doctor is on the way to complete amnesia.[1]

The victim of this memory meltdown is the U.S.S. *Voyager's* only physician, a computer-generated Emergency Medical Hologram (EMH) in human form (VOY-3, "The Swarm"). The diagnosis is simple: the EMH's memory has become overloaded. This virtual doctor was designed to be a temporary, emergency backup to the ship's real physician. The skills and knowledge of many physicians were stored in the ship's computer, to be available in case the real physician was ever temporarily incapacitated. But the ship's real physician has been killed, and the EMH has been on duty for most of two years. During that time, the virtual doctor has diagnosed and treated numerous medical problems, just as he was designed to do. These medical activities don't push the limits of his memory. But he's also been moonlighting—reading, enjoying music, establishing relationships. Doing all these things that his designers never intended him to do has seriously overloaded the limited capacity of his memory circuits.

The only available solution is a drastic one—the EMH's memory must be wiped clean. This is the kind of reinitialization that you do when you reformat a computer disk—everything is wiped clean. While *Voyager's* officers debate whether to reinitialize their holographic crewmate, Kes, the EMH's nurse and close friend, points out the personal toll that reinitialization would exact. All those things that the EMH had learned or experienced would vanish: his friendships, his passion for opera, his sense that he is a valued crew member—even his memory of having fallen in love. All would disappear with a single push of a button.

Kes's appeal on behalf of her holographic friend shows something important: memory literally defines who we are. Our values, our attitudes, our habits are all characteristics that are unique to each of us. And all of these characteristics form part of our memories. Memories help us recognize our friends and loved ones, and they allow us to anticipate the future by learning from the past. Memories provide a transcript of an individual's life history and a guide for planning the future. To erase your memory is to erase your sense of self.

A computer-generated hologram's memory is not the same as a human being's. For starters, it has a limited capacity. The holodoc's overabundance of experiences has overloaded his circuits. But there's no evidence that human memory has an upper limit. Moreover, the holodoc's memory is stored in the electronic circuits of a computer; human memory resides within biological circuits made of neurons and supporting tissue. Despite these differences in capacity and hardware, however, the virtual doctor's loss of memory has something in common with an all-too-real human condition, Alzheimer's disease (AD).[2] Today, about 4,000,000 people in the United States alone suffer symptoms of AD, including difficulty concentrating and remembering.

The holodoctor's fictional memory loss brings pain both to the victim and to those around him. His close friend Kes is devastated when he cannot remember who she is. AD does the same thing, creating a thick, impenetrable wall between victims of the disease and their loved ones and friends. Even though Alzheimer's disease is not a matter of overloaded memory circuits, the parallels between the holodoc's condition and the effects of this disease are striking.

Early on, the holodoc tries to deny his incapacities, fabricating all sorts of explanations for his forgetfulness—distraction, fatigue, anything but the hard truth that his memory is disintegrating. This same denial characterizes the early stages of AD. But as the disease progresses, excuses give way to desperation. At this point, a caregiver can make things easier for the AD victim by breaking complex tasks into simple steps, allowing the sufferer to complete one step at a time without having to keep the whole task in mind. This is exactly what Kes did as she patiently prompted the holodoctor step by step through the operation.

Disorderly Conduction

While a cure for AD is not on the immediate horizon, we do know something about how the disease progressively alters the human brain. Neurons in the brain do their work by "talking" to one another, exchanging and modifying information. The currency of this information exchange is a collection of chemical messengers called neurotransmitters. Proper neural function—and hence normal cognitive function—depends upon the neurons' ability to send and receive these neural messages. For the messages to make sense, they must originate from the correct neurons and must be transmitted to the appropriate target neurons. Precise, coordinated teamwork among neurons is possible only if the terminal endings of those neurons are in close proximity. In a healthy brain, neurons are interdigitated like the branches of closely spaced trees. Anything that distorts these intricate patterns of branching, or reduces the number of branches, disrupts the proper flow of neural messages. When neural messages get scrambled or misrouted, neural chaos and behavioral dysfunction result. And this is exactly what happens in AD.

AD mounts a two-pronged attack on the brain's nerve cells, with one prong aimed at the outside of the cells and the other focused on the inside. Together, these attacks induce the growth of pathological structures in and around nerve cells. These abnormal structures can easily be seen in magnified pictures of brain tissue. Look at the pair of photographs on the facing page. The one on the left shows a magnified view of neural tissue from a healthy brain. The dark "stripes" with "bulges" are neurons arrayed in neat, orderly fashion;

the bulges are the cell bodies of these healthy brain cells. The photograph on the right shows tissue from the brain of a patient with AD. Here, the neurons are disordered. Their parts form a mass of "tangles," and the blurry dark regions correspond to the abnormal buildup of globs of protein called "plaques."

Plaques and tangles work in tandem to do a number on the neural circuitry needed for normal brain function. Like plaque that builds up on the teeth, neural plaques are deposits or coatings that build up on the surfaces of neurons. In the case of AD, the offending plaques consist mainly of a substance called beta-amyloid protein (or BAP). When a glob of BAP is deposited on the outside of a neuron, that neuron swells and becomes twisted and misshapen.

That's the bad news on the outside of the cell. But there's more bad news on the inside. Within a healthy neuron are long, thin structures called *fibers* or *filaments*. These structures are tiny intracellular highways that carry chemical materials, including nutrients, from the cell body, where they are manufactured, through the branches to the neuron's outlying terminals, where they are needed to create the electrochemical signals that the neuron uses to communicate with other neurons. Tangles are filaments that have become abnormally twisted and broken. As these molecular potholes build up on the intracellular highways, nutrient transport is disrupted, and the neuron's all-important

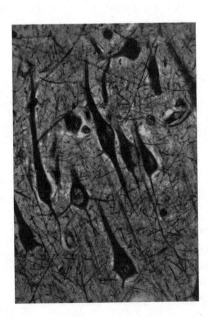
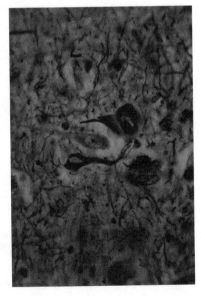

terminals are starved. Eventually, the outlying branches degener-
ate and disappear. As these structural changes on the inside and
outside of neurons progress, memory and other cognitive functions deteriorate.

"It's a Horrible Feeling Not Knowing Who You Are, Where You're From."[3]

Voyager's officers decide to reinitialize the holodoctor's failing
memory, but this cure turns out to be almost as bad as the disease.
The doctor's memory of the two years he's been on board *Voyager* is
wiped clean. Suddenly all the ship's crew are complete strangers to
him. As his former friends are mourning the lost relationship,
something strange happens: their memoryless friend spontaneous-
ly breaks into quiet song, a Puccini aria he learned before the reinitialization. This reveals that not all of his memories were
destroyed by the reinitialization. Inexplicably, some of the doctor's
Voyager experiences have survived the therapy.

It seems odd that the holodoctor can sing an aria, but not recognize his colleagues or remember anything about his past. How
can an individual lose some parts of his memory while retaining
other parts? Puzzling as it may sound, this is precisely the kind of
selective memory loss that often occurs in humans. In fact, selective memory loss provides a key insight into the nature of memory and how it is organized within the brain.

Selective memory loss shows that what we call "memory" is
not a single, unitary process. Instead, we have several types of
memory, each working more or less independently of the others.
Normally we're unaware of this division of labor, but it is thrown
into sharp relief when certain portions of the brain are damaged.
This is what happened in the famous case of a twenty-seven-year-
old human male whom the medical literature calls H. M. (out of
respect for his privacy).

In Connecticut in the summer of 1953, a neurosurgeon
attempted to relieve H. M.'s intractable epilepsy. Believing that
portions of the temporal lobe were responsible for generating
epilepsy's electrical storms in H. M.'s brain, the surgeon removed
those portions of the brain, which included the structure known
as the hippocampus. The surgeon had done similar operations on
other patients before, but the effects on H. M. were different,
unexpected—and catastrophic.

> . . . [I]mmediately after Mr. M. left the recovery room it was
> clear that something else was dreadfully wrong. He was
> uncomprehending. Yes, he could speak and read and
> converse. But when asked where he was, and who were
> these people at his bedside, he did not know. He could not
> find his way to the bathroom, and nurses could enter the
> room, speak to him, leave, and then return a moment later
> to find he had no memory of them.[4]

Today, more than four decades later, H. M.'s ability to form new memories has not recovered. It never will. He suffers profound anterograde amnesia: he can recall things learned before the surgery, but finds it utterly impossible to remember new facts or events for longer than a minute or so. Right after eating breakfast, H. M. can tell you what he had, but ask him thirty minutes later and he won't even remember *that* he ate, let alone *what* he ate. H. M. can read a magazine, put it down, and within a few minutes forget having read anything. He's liable to pick up the same magazine and read it again and again, each time as though it were new to him. He is unable to remember the names of people he's met since his surgery, although his memory for names and faces encountered prior to surgery remains intact. H. M. needs maps to get around his neighborhood, and he needs to be reminded to remember to use those maps.

But here's where the story really gets interesting, and where H. M. begins to resemble *Voyager's* holodoc: he isn't totally incapable of forming new memories. He can't learn and remember the name of street he lives on, but he can learn and remember new skills—although he doesn't remember that he's learned them. For example, H. M. learned to solve the Tower of Hanoi, a challenging puzzle that involves moving rings back and forth among a series of pegs. With one day's practice, H. M. got pretty good at the puzzle. The next day, when asked about the puzzle, he had no recollection of ever seeing it before. Yet when he sat down to try it, H. M. did much better than you'd expect from a complete novice— he remembered how to solve a puzzle that he didn't remember learning. His improved performance, which has endured, reveals the existence of a kind of memory that's different from the memory used to recall facts and names. Neuropsychologists use the term *memory dissociation* to refer to the existence of such specialized types of memory that can function independently of one another.

Evidence for memory dissociation is also seen in victims of AD, at least during the early stages of the disease. As AD progresses, the damaging tangles and plaques accumulate more rapidly in some parts of the brain than in others. Why certain parts of the brain are more susceptible to AD remains an important unsolved mystery, but whatever the reason, this selective degeneration produces character-istic cognitive deficits.[5] As a rule, during the early stages of the dis-ease, patients forget facts such as the items they're supposed to buy at the grocery store, but they do not forget skills, such as how to tie their shoes.[6] These selective losses can be maddening for the patient. Just imagine having no trouble driving your car—an amazingly complex skill—yet forgetting your destination. One of the early, frightening symptoms of AD is disorientation: AD victims often experience trouble finding their way around, even in familiar places.

During the later stages of AD, the neural degeneration spreads to areas of the brain initially spared. At this stage, the situation becomes grave. Memory for learned skills also begins to deteriorate, rendering the patient truly and profoundly demented. As disturbing as H. M.'s condition is, he might be thankful for that portion of his memory that remains, if he could only remember that it was spared!

We have names for the two types of memory implicated in AD and in cases of amnesia, including H. M.'s: declarative memory and procedural memory.[7] Declarative memory is a vast repository of conceptual and factual knowledge. As its name implies, declara-tive memory allows you to declare information, to yourself and to others. Declarative memories can be brought to consciousness in response to a question. When you answer "Boston" to the question "What is the capital of Massachusetts?" you draw on declarative memory. Ditto when you tell someone that the Civil War ended in 1865. When you tell a friend what you ate for breakfast (and only a friend could care), that autobiographical trivia comes from declar-ative memory. And when you can't remember the name of an acquaintance, it's declarative memory that has failed you.

Procedural memory is quite different. It comes into play when you perform learned skills or reflexively react to previously encountered objects. When you play the piano, or when H. M. does the Tower of Hanoi puzzle, it's procedural memory that's being used. You don't have to consciously recall procedural mem-ories—they are often triggered automatically and can guide your actions without conscious effort.

Some test questions may help to clarify the difference between these two types of memory. What is your personal identification number for your ATM card? Some people can rattle off their PIN readily—they're using declarative memory. But for others, the answer comes only by pretending to punch in the numbers on an imaginary keypad. Procedural memory, in other words, may hold the answer even though declarative memory may not.

Here are a couple more questions. When putting your pants on in the morning, which leg do you start with? Which way do you turn the knob on your stove to light a burner? Answers to such questions often require the involvement of procedural memory, which guides us through those actions when we actually execute them. On Deep Space Nine, procedural memory serves Chief Miles O'Brien when he's lying in a cramped Jeffries tube rewiring a blown circuit. Likewise, procedural memory is what Sirna Kolrami has developed exquisitely to become a grand master at strategema, a three-dimensional game that demands speed and dexterity (TNG-2, "Peak Performance"). And on the bridge of the *Enterprise*-D, those ensigns who obediently implement Captain Picard's orders are relying on procedural memory as their fingers fly across the touch-sensitive console.

We see the distinction between declarative memory and procedural memory in a *Star Trek: The Next Generation* episode in which everyone aboard the *Enterprise*-D develops selective amnesia (TNG-5, "Conundrum"). An energy beam from an alien vessel sweeps the *Enterprise*-D, and the entire crew on the bridge is stunned by the consequences. No one can recognize anyone else—they're all complete strangers to one another, and they have no idea where they are or what they're up to. "I don't even remember who I am," laments Captain Picard. Their declarative memories have been wiped clean—they have retrograde amnesia.

Despite this loss of memory, no one panics. Each crew member tackles some chore—Ensign Ro operates the pilot controls, Worf punches in numbers on the tactical control panel, Dr. Crusher reflexively uses the medical scanner to carry on her duties in sick bay, and Riker picks up his trombone and plays a tune, even though he has no memory of ever handling the instrument before. Clearly, their procedural memories have survived the assault on their brains.

The *Enterprise*-D's main computer has suffered selective amnesia, too. All the crew's personnel records have been erased, but the rest of the data files remain intact. Chief Engineer LaForge recognizes that this selective loss of records is suspicious, and Commander Riker and Dr. Crusher agree.

RIKER: It's a little too selective to be a coincidence.

CRUSHER: As selective as what was done to our own memories. Skills are still in place, but personal knowledge is unavailable.

By studying previous medical cases, Dr. Crusher comes up with a treatment that reinstates the lost memories. According to her, "It [the treatment] involves increasing the activity of the medial temporal region of the brain. Using short-term memory synapses to retrieve long-term memory." Although this treatment is unconventional by present-day medical standards, it succeeds in restoring the *Enterprise*-D's crew to normal.

Procedural memories—the sort that the amnesic crew held onto—work perfectly well without the fuss and bother of conscious deliberation, as when H. M. learned the Tower of Hanoi puzzle without even being aware that he had done so. The same is true of many of the memories that we form in our everyday lives. During the course of a day, we encounter people and objects that are of no immediate importance to us. Consequently, we have no explicit intention of remembering anything about those people or objects. And yet our brains register at least some details of those encounters, making it easier to recognize those people or objects when we see them again.

The formation of such implicit memories is nicely demonstrated in a pair of feature-length *Star Trek* movies. To set the stage, at the end of the second *Star Trek* movie (*Star Trek II: The Wrath of Khan*), Mr. Spock makes emergency repairs to the *Enterprise's* core, but in so doing exposes himself to lethal radiation. He sacrifices his life for the rest of the crew. As he lies dying, Spock gives a typical Vulcan explanation for his selfless act. Logically, he says, "the needs of the many outweigh the needs of the one." In the third *Star Trek* movie (*Star Trek III: The Search for Spock*), Federation science and Vulcan religion combine forces to resurrect Spock, but they cannot restore his memory to its pre-death condition. The resurrected Spock walks right past his former crewmates without recognizing them, though he senses something vaguely familiar about one of them, his old friend Captain James T. Kirk. Explaining to Spock why the *Enterprise* came back to rescue him, Kirk says, "because the needs of the one outweigh the needs of the many." The phrase primes Spock's implicit memory of who Kirk is. This rekindled memory allows him to say, "Your name . . . is Jim."

Note that Kirk's comment does not repeat verbatim what Spock himself had said, but the similarity is enough to establish a link in Spock's memory. This incident illustrates one of memory's most valuable attributes. Rarely does an entire event repeat itself perfectly—but it doesn't have to. We can recollect an event from just a few fragments of memory associated with it. This phenomenon is termed *cued recall*, and it's an indispensable characteristic of human memory.

In another *Star Trek: The Next Generation* episode, cued recall literally saves the *Enterprise*-D from destruction (TNG-5, "Hero Worship"). A young boy named Timothy has been brought on board the *Enterprise*-D, the only survivor of the mysterious explosion of a Federation science vessel, the S.S. *Vico*. Traumatized by the loss of his family and friends, Timothy cannot remember what triggered the explosion. Captain Picard and Counselor Troi plead and cajole, but it's no use: the secret is locked up in Timothy's head. Timothy's amnesia is particularly unfortunate, because the *Enterprise*-D is experiencing exactly the same kind of gravitational wave fronts that destroyed the *Vico*. Alarmed by the growing gravitational force, the *Enterprise*-D crew frantically discusses options, and this cues, or primes, Timothy's memory:

> RIKER: Riker to LaForge—can you give me more power to the shields?
>
> LAFORGE: Stand by.
>
> TIMOTHY: [to Data, with excitement in his voice] That's what they kept saying, "more shields, more shields."
>
> DATA: [to Timothy] I want you to recall everything you heard people say aboard the *Vico* before it was destroyed.
>
> TIMOTHY: I don't know . . . just that, just "more shields."
>
> LAFORGE: [after some discussion of options] Diverting warp power to the shields.
>
> TIMOTHY: [to Data] "Warp power to the shields"—they said that too, Data, I'm positive.

In the nick of time, Lieutenant Commander Data realizes that the ship's own shields are responsible for the amplified gravitational force. Data convinces Picard to drop the shields, and the *Enterprise*-D is saved. None of this would have been possible if cued recall had not awakened Timothy's memories of similar events on the ill-fated *Vico*.

A Tale of Two Systems

In the absence of brain damage, the two dissociated memory systems—declarative and procedural—operate simultaneously, each carrying out its own job for the common good of their owner. These two memory systems can coexist, in part, because they're are headquartered in different parts of the brain.

The circuits that make it possible for you to learn and remember are widely distributed throughout the entire brain. Still, among all the brain's billions of neurons, two groups of neural structures have special importance, one for each memory system. One group has the main responsibility for declarative memory. This group is called the *hippocampal system*, or *medial temporal lobe system*, in recognition of the prominent role of the hippocampus and other temporal lobe structures. The principal elements of this system are shown in the illustration below. It was surgical destruction of key components of the medial temporal lobe system that robbed H. M. of the ability to form new declarative memories.

Procedural memory depends upon a different set of brain structures, called the *neocortical processing system*. This system includes evolutionarily newer regions of the cerebral cortex (*neo* means "new"). The term *neocortex* encompasses most of the cerebral cortex, including its sensory areas, but excluding the evolutionarily older regions that process smell information.

Although there is some dispute about the neocortical processing system's precise boundaries, everyone agrees that it includes several large groups of neurons that lie outside the neocortex itself. Collectively, these groups make up the *basal ganglia* (*basal* refers to their location near the base of the brain; *ganglia* means

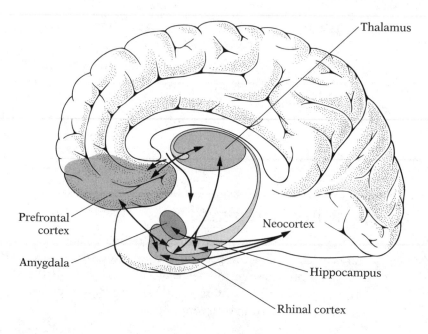

"large groups of neurons"). The basal ganglia are the core of the circuitry that's responsible for coordinating sensory inputs and motor behaviors. As the illustration below shows, the basal ganglia receive most of their *input* signals from the sensory systems and send their *output* signals to motor areas in the frontal cortex. Thus these structures are crucial for the performance of various motor skills, such as driving a car. Because H. M.'s surgery spared these regions of his brain, his procedural memory system remained intact. Ironically, he could learn and retain all sorts of new motor skills, even if he couldn't learn a new home address.

In normal, healthy individuals, these two systems work side by side, and we're oblivious to their distinct operations. Their divided responsibilities and their cooperation are invisible to us. Damage to the brain can disrupt this lovely harmony, revealing memory for what it really is: two separate systems.[8]

The Long and the Short of Memory

Commander James Kirk is being court-martialed for the death of one of the *Enterprise's* crew members (TOS-1, "Court Martial"). During the proceedings, Kirk identifies himself for the court record, reciting his Starfleet serial number: SC 937-0176 CEC.

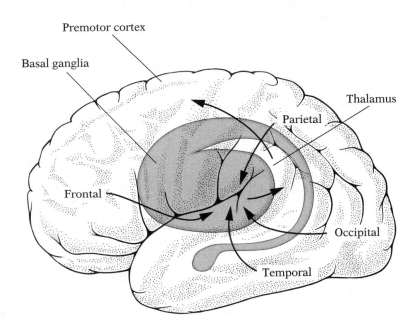

Like most long, meaningless strings of digits and letters, Kirk's serial number is hard to remember. You've just read the number, but can you repeat it *without* looking back to refresh your memory? Right now, you may be able to repeat all of the number, or least much of it, with hardly any effort. But if you try to recall the number a few hours later, you'll probably discover that it has been erased from your memory. The immediate, effortless, ephemeral memory that allowed you to hold onto the number for a few seconds is aptly called *short-term memory* (STM), and is distinguished from its longer-lived relative, *long-term memory* (LTM).

As you walk along the beach near the ocean's edge, each step you take leaves its mark in the sand, but those marks last only a short time. Someone else's footprints obliterate yours, or the ocean's waves wipe the record clean. Likewise, our everyday experiences—people, conversations, events—leave their marks on our short-term memory. These footprints in the neural synaptic connections of our brains hold information about those recent experiences. That telephone number you briefly remembered was held in STM, and the neural footprints associated with it allowed you to remember it long enough to make a call.

STM's neural footprints don't last forever, but they can help you wiggle out of some embarrassing situations. Someone is talking to you, but your mind has wandered off. Realizing that you're daydreaming, the speaker accuses you of not listening. You reach into your STM and repeat the last few words the speaker uttered, thereby "proving" that you were listening.[9]

STM's contents are impermanent. Unless they are moved into more permanent storage, their decay is inevitable. In some cases, one moment's STM contents are replaced by the next moment's contents. In other words, STM is overwritten, like a drawing on an Etch A Sketch. A phone number you're trying to remember will be erased from your STM if someone gives you a new one to keep in mind. STM, in other words, has very limited storage capacity. And even if the contents of STM are not overwritten, natural decay wipes them away within a few minutes or so, once they've outlived their usefulness.

This kind of brief, "scratch pad" memory is very useful for immediate projects, such as carrying on conversations and dialing phone numbers. But STM cannot come close to handling all the memory needs of organisms whose life spans are measured in decades. Many of our actions and decisions must be based on knowledge of events that happened days, weeks, or even years ago.

Here's where LTM comes in. The distinction between STM and LTM resembles the difference between a computer's temporary, random access memory (RAM) and its permanent, read-only memory (ROM) or hard disk. Information stored in RAM is lost when the computer is powered down, but information in ROM or on the hard disk will still be there when the machine is turned on again.

Of course, it's not possible to power down and then restart a brain to see what memories are retained. The closest we can come to a temporary powerdown is a severe bump on the head. When the bump is sufficiently hard, the resulting trauma to the brain can make the victim forget everything that happened immediately prior to the accident. This is exactly what happens when Quark, the Ferengi owner of the bar on Deep Space Nine, gives his brother Rom and nephew Nog a lift to Earth on his new personal shuttlecraft (DS9-4, "Little Green Men"). Ostensibly, Quark is being a good guy, taking Nog to Earth so he can attend Starfleet Academy. (Nog is the first Ferengi ever admitted.) But, as usual, Quark has an ulterior motive—he's smuggling contraband. As Quark's shuttlecraft approaches Earth, it crashes.

When the three Ferengi awaken, they are in a U.S. Army facility in Roswell, New Mexico, and the year is 1947. Quark and his fellow travelers have no idea what has happened to them. They remember thinking that they were about to crash, but cannot remember the crash itself. The events of the crash did get into their short-term memories. But the crash—and bumps on their heads—prevented the contents of their short-term memories from being transferred into more permanent storage.

In humans, the hippocampus plays a key role in the transfer of information from STM to LTM. We know this because people with damage to the hippocampus, including patient H. M., can form perfectly normal short-term memories. What they can't do is transfer the contents of STM into permanent storage.[10] It's as if they are getting an unending series of memory-disrupting bumps on the head.

Take Two Pills and Remember It All in the Morning

Lieutenant Reg Barclay, to put it kindly, is not exactly the Albert Einstein of the *Enterprise*-D's engineering section. When someone asks him a question, Reg stammers out some unimaginative answer; when he plays the lead role in a shipboard theatrical production, he blows his lines. He tries hard, but frankly, by Starfleet standards, he's a bit dim. But all this changes when the *Enterprise*-D investigates the malfunction of the huge Argus Array subspace telescope (TNG-4, "The Nth Degree"). An alien probe is spotted lurking near the Array, and Geordi LaForge is assigned to investigate. Being a nice guy, LaForge asks Reg to accompany him. As their shuttlecraft approaches and assays the intruder, the probe emits an intense pulse of energy, which knocks Barclay down.

One morning shortly after this incident, Geordi comes upon Reg in the holodeck. Reg is having an animated conversation with a computer-generated Albert Einstein, and together they have filled blackboard after blackboard with indecipherable equations. Geordi asks what's going on. After all, quantum electrodynamic calculations are normally not Reg's strong suit. Einstein's intellectual sparring partner begs to differ, casually asserting that he's just never thought along these lines before. In fact, Reg says with self-satisfaction, it's all pretty simple stuff, if you put your mind to it. Geordi suspects that there's more going on than Barclay's putting his mind to it, and that the probe is somehow responsible.

It turns out that Geordi is right. The pulse of energy from the probe has initiated a series of changes in Barclay's brain, and these changes have turned him into a very bright bulb indeed. When Dr. Crusher examines the pattern of activity in Barclay's brain, she is astonished at the accelerated production of neurotransmitters, those brain chemicals that carry messages among neurons. She is equally amazed by the structural changes in the membranes around Barclay's neurons, changes that heighten the neurotransmitters' effectiveness still more. No wonder, then, that Barclay's IQ is off the scale, supposedly in the neighborhood of 1,200 to 1,500. These changes bless Barclay with an awesome intellect and an absolutely phenomenal memory, as we see when he performs flawlessly in the title role of *Cyrano de Bergerac*.

Barclay's picture-perfect memory reminds us that the creation of stable, enduring memories requires minute anatomical changes in our neurons. These changes include the formation of additional neuronal branches, and thus the formation of more synaptic connections per neuron. On future occasions, neural signals tend to follow these newly strengthened links. This situation is analogous to the growth of highways and streets in a developing suburb. Existing roads are widened, and smaller side streets grow into the expanding neighborhood, branching off the existing traffic arteries. In theory, all these new roads make traffic flow more efficiently.

The layout and maintenance of new roads depends on builders and city planners, but the production and maintenance of new neural circuits depends on DNA. Genes control the production of the proteins that are used to build and activate new neural circuits. This knowledge suggests a straightforward approach if an individual's brain doesn't have the pathways that are normal for its species.

> *I was six, small for my age. A bit awkward physically, not very bright. In the first grade when the other children were learning to read and write and use the computer, I was still trying to tell a dog from a cat, a tree from a house. I didn't really understand what was happening. I knew that I wasn't doing as well as my classmates. There were so many concepts that they took for granted that I couldn't begin to master. And I didn't know why. All I knew is that I was a great disappointment to my parents.*
>
> Dr. Julian Bashir (DS9-5, "Doctor Bashir, I Presume")

How did one of Deep Space Nine's most intelligent crew members get transformed from a dullard into a genius? Around age seven, he underwent a procedure called "accelerated critical neural pathway formation." Careful manipulation of his DNA promoted the growth of neural networks in his brain—precisely the sort of changes that might make it easier for him to form crucial long-term memories.

Long-term memories depend upon the genes that produce a particular class of proteins known as CREB (for *c*yclic adenosine monophosphate *r*esponse *e*lement *b*inding proteins). These memory-building proteins are synthesized within the neurons themselves, and they do their work by turning on and turning off other genes that are responsible for the creation of new synaptic branches and connections.[11] When those genes are turned on, they do what any gene does: produce the protein specified by that gene. One member of the CREB family turns off the memory-building genes and thus *represses*, or blocks, the formation of long-term memories. An animal whose memory circuits are flooded with CREB's repressor form will have perfectly normal STM, but little ability to convert it into the more enduring LTM form. Thus, the hapless creature may remember something perfectly well for a few moments, only to have the memory dissolve in half an hour or so.

A complementary form of CREB *promotes* the formation of long-term memories. Suppose it ordinarily takes at least ten practice trials for an animal to learn how to negotiate a maze to obtain a food reward. An animal whose brain contains an abundance of CREB's promoter form may be able to master the task in just two or three trials.[12] The CREB promoter has turned a struggling learner into a genius.

Before you rush out to buy a CREB-based wonder drug to turbocharge your memory, you should keep a few things in mind. Even if such a drug were available right now—which it isn't—it would have risks. Normally, the vast bulk of the information entering STM never makes it into LTM. And this selective entry into LTM makes good sense. LTM's resources should be reserved for experiences of importance to us, memories that we'll need in the future to guide our actions. Imagine what would happen if, thanks to CREB-like drugs, all your experiences, no matter how inconsequential, found their way into LTM. Memory can be too good for its own good, as Lieutenant Commander Data shows on occasion.

Too Much of a Good Thing

Data certainly doesn't need memory-improving drugs or the sort of treatment that young Julian Bashir got. Data's positronic brain endows him with a very large, seemingly perfect memory. But that memory occasionally causes trouble when a humanoid crewmate asks Data to recollect some fact or event. Here's Data analyzing a dangerous situation that's coming to a head (TNG-1, "Justice"):

> DATA: It was probably unwise of us to attempt to place a human colony in this area. There are 3,004 other planets in this star cluster that we could have colonized. The largest and closest . . .
>
> PICARD: Data! Don't babble!
>
> DATA: Babble, sir? I am not aware that I ever babble, sir. It may be that from time to time I have considerable information to communicate. And you may question the way in which I organize it . . .
>
> PICARD: Please . . . organize it into brief answers to my questions. We have very little time.

When Data accesses his enormous store of memories, he tends to go on and on, dredging up every last item remotely relevant to the query. From the asker's point of view, this is overkill.

The lesson? The usefulness of memory depends on more than just the volume of material remembered—we must be able to extract, or recall, just what is needed, no less and no more. LTM in humans is organized to promote efficient retrieval, and this organization is assisted by the selective weeding of STM's contents.[13]

Alexander R. Luria, the most influential Russian neuropsychologist of the twentieth century, described a human—nicknamed "S."—who was the near-equivalent of Lieutenant Commander Data. Luria was so fascinated by this man that he studied him for fifty years, eventually summarizing his findings in a delightful little book, *The Mind of the Mnemonist*. Here's just one example of S.'s powerful memory. Luria read to S. a list of meaningless digits that had no pattern to their order—the length of the list exceeded one hundred digits. Years later, S. could still repeat the entire list from memory, one digit after another, without error! In contrast, a human with normal intelligence can have trouble remembering more than seven or eight unrelated digits.

Don't be depressed because you don't have S.'s perfect memory — almost no one does. And anyway, a perfect memory doesn't guarantee brilliance—or happiness, for that matter. Courtesy of his gargantuan memory, S. could read paragraph after paragraph of text and then recall everything he'd read, word for word. But, as Luria wrote,

> *S.'s grasp of entire passages in a text was far from good. . . . [O]n first acquaintance, S. struck one as a disorganized and rather dull-witted person, an impression that was even more marked whenever he had to deal with a story that had been read to him. If the story was read at a fairly rapid pace, S.'s face would register confusion and finally utter bewilderment. "No," he would say. "This is too much. Each word calls up images; they collide with one another, and the result is chaos. I can't make anything out of this. And, then, there's also your voice . . . another blur . . . then everything's muddled."*[14]

Boosting Your Own Memory

Unlike Data and S., most of us get by perfectly well with limited, judiciously selected information in our LTM. Moreover, we're able to deploy special strategies for accessing that information. Here's an example of one such strategy. Imagine being caught without a calendar but needing to know the total number of days in the month of April. You would probably recite the tried and true mnemonic, "Thirty days hath September, April, June, and November . . . " This mnemonic exploits rhyme and rhythm to categorize information, thereby making it easy to recall. Medical students rely on similar memory aids to learn lists of anatomical features, such as the names of the twelve cranial nerves.

Another useful strategy for creating and retrieving memories relies on our keen ability to form visual images. The idea is simple: generate an image of some familiar object or place, and then associate that image with the new item you need to remember. Waiters in restaurants use this device all the time to remember food orders. They form a mental caricature of each diner's face and then morph that caricature into the shape of the food object ordered by that person. Think about this strategy the next time a waiter carefully scrutinizes your nose as you order lobster.

A related version of this imagery strategy exploits the rooms of a familiar building to provide memory-enhancing mental pegs. Imagine you're headed to the grocery store and must remember to purchase seven items for dinner. Select a room in your house, such as the living room, and visualize the furniture in that room. Now mentally walk around that room and "place" a food item to be purchased on each piece of furniture—potatoes on the couch, coffee on the coffee table, flank steak on the CD player, and so on.[15] Now simply go to the grocery store and repeat your mental tour of the living room—each familiar piece of furniture will remind you of its associated food item. This simple but effective strategy has been known since the Roman orator and statesman Cicero let people in on the secret in 55 B.C.

After reviewing the memory improvement methods that had been proposed over several centuries, William James concluded that "All improvement of memory consists, then, in the improvement of one's habitual methods of recording facts."[16] What James called "recording facts" we now call "encoding of information." But James's basic point remains sound. To understand what encoding means, let's think about words. Any word can be encoded at any of several different levels of complexity, depending on the demands of the task. The following simple exercise illustrates this point. On the facing page there is a group of words within a box. Each word is printed in one of two fonts, either **Chicago** or Helvetica. Your task is to name the font each word appears in, while working your way through the words as quickly as possible.

Don't actually pronounce each word, just name its font (**Chicago** or Helvetica). When you're done, come back to this sentence— *do not read the next paragraph just yet.*

---

elephant border leaves **jingle** trumpet **cause damage belt** glory **nurse** glove alley **smart tingle contract** simple **banana** foil drift **seize garage** stem syringe **gamble** float **derivative** pudding

---

Now that you've completed the font-naming task, let's see how many of the words you can recall. Jot down the words you remember, and check your accuracy. If you're like most people, it's not so good. But that's to be expected. When you named the fonts, you didn't attend to the meanings of the words themselves—your encoding process skimmed along at a relatively shallow level of analysis. Your recall would have been much better if instead of naming the fonts, you had had to think up a synonym for each word. Then each word would have been more deeply etched on your memory.

Perhaps you're one of the many people who have trouble "remembering" people's names after being introduced to them. This, too, could be related to levels of encoding. Being visual creatures, we tend to pay attention to the facial characteristics of a new acquaintance as we're being introduced. Consequently, the verbal introduction—"Hi, I'm Elaine"—tends to go in one ear and out the other. But the solution is simple: force yourself to process the name itself at a deeper level of analysis. Think of other names that rhyme with it. Associate the name with some unique facial characteristic. Imagine the name tattooed on the person's forehead. Anything that encourages you to encode the name itself will help stamp it into your memory.

Depth of encoding also depends on the importance of the material you're trying to learn. Here's an example of this principle. Ferengi attach great importance to their aphoristic guides to a successful life in Ferengi society. Every Ferengi child must memorize the nearly three hundred principles codified in the Ferengi Rules of Acquisition. If asked to state the 21st Rule, the child is expected to shoot back unhesitatingly, "Never place friendship above profit" (DS9-2, "Rules of Acquisition"). When asked for the 49th Rule, a Ferengi child must reply, "The bigger the smile, the sharper the knife." Just how well learned are these rules? During an away mission, Neelix—a Talaxian and *Voyager's* cook—is required to masquerade as a Ferengi so he can force two genuine Ferengi to give up their lucrative stranglehold on the citizens of a backward planet (VOY-2, "False Profits"). In an attempt to impress the Ferengi, Neelix quotes a plausible but bogus Rule of Acquisition: "Rule 299: 'Whenever you exploit someone, it never hurts to thank them.'" The two Ferengi are immediately suspicious, for there are only 285 rules, and like all Ferengi, they know each and every one of them. For Neelix, however, these rules are not important, so he's never bothered to learn them. As result, the Ferengi quickly uncover Neelix's ruse.

Ordinary people can have quite extraordinary memory abilities, depending on what information they *need* to remember and how they encode that information. Before the age of automatic teller machines, human bank tellers often knew at a glance the faces and names of several hundred regular customers. Even today, experienced sports announcers can recall decades' worth of statistical data about games and players. Some musicians and conductors are renowned for their ability to play or direct literally thousands of measures of music entirely from memory. And accomplished card players can remember what cards have been dealt during a hand, which allows them to calculate the changing odds. Lieutenant Commander Data routinely counts cards when he plays poker with his fellow crew members (TNG-2, "The Measure of a Man").

Putting Memory to Work

When you were six years old, how many windows did your bedroom have? Or try this: Add up all the individual digits in the street address where you were living at that age. Now think of a English word that rhymes with the name of your street.

When you struggle with these mental tasks, you're working with memories right at the fingertips of your mind. Much of what we do with memories entails manipulating material that resides temporarily in this *working memory*.[17] You can bring up information from long-term memory into working memory, where it is readily accessible. To come up with the answer to the first question, you probably had to recall where you lived when you were six and recreate the layout of your bedroom before you could mentally count the windows. Typically we have discretionary control over the contents of working memory, and we determine how long information remains there. (How disturbing it would be for a thought in working memory to persist beyond its usefulness! What if, day in, day out, year after year, you couldn't stop thinking about your childhood bedroom?)

Brain imaging studies tell us that working memory's discretionary control is carried out by neural circuits in the prefrontal cortex, a region located toward the front of the brain. Some neurons in this relatively large brain area become active when a person tries to hold in memory a short series of consonants or images of human faces. This heightened activity is maintained for many seconds, coinciding with the length of time before the person is asked a question about those consonants or faces.[18]

Damage to the prefrontal cortex undermines the integrity of the circuits that control working memory. In particular, the contents of LTM, while unaffected themselves, are liable to pop willy-nilly into working memory. Aging is one condition associated with impairment of circuits in the frontal lobes, the area of the brain that houses the prefrontal cortex. As a result, some elderly people have difficulty screening material out of their working memories, however inappropriate such material may be to their current situation.[19] For example, here's how two elderly individuals responded to the question, "How much education did you get?"

PERSON 1: High school and then a bit of college.

PERSON 2: Well, let's see. I went to school in _____ where, uh, uh, I grew up in _____. Back in those days, why they didn't have the big high schools they have now. When I went back there a few years ago in . . . uh, I don't remember exactly when it was. I think it was the summer of 1980 or maybe it was 1981. I went for my brother's 50th anniversary, and I didn't recognize the place at all. We went to a small school, the only school in town. It was the only place to go. All the children were in one room. The school only went to Grade 9 or uh, uh, I think was . . . was it Grade 9? No, it was only Grade 8 because _____ (neighbor's daughter) left to go to nursing school and she had to go to _____ finish Grade 9. She never finished nursing anyway. She got married but it didn't last long.[20]

For people like the second individual, long-term memory remains normal, but its contents leak, unbidden, into working memory and then out onto the tongue. These examples represent extremes in leakage, and obviously, aging doesn't affect everyone in exactly the same way. These individual differences probably reflect differences in the way that aging affects the brain.

How Much Data Can Data Remember?

In all of Starfleet, no brain is more famous or more coveted than Lieutenant Commander Data's. Its memory is a lightning-fast, high-fidelity device with enormous capacity. When asked how large his memory is, Data gives a characteristically precise answer: 800 quadrillion bits (TNG-2, "The Measure of a Man"). If you dumped the entire capacity of Data's memory onto ordinary 1-megabyte floppy computer disks, you'd need 100 billion of them. And if you stacked those floppies one atop the next, the resulting pile would be an air traffic-menacing 197,285 miles high.[21]

But even this stack of floppies couldn't handle all the events of an ordinary human life. Your eyes all by themselves would send the brain enough information to use up all of Data's memory storage capacity, and then some.[22] And this wouldn't leave any room for the information streaming in from the other senses, or from the process of thought itself.

To deal with this torrent of information, the human nervous system takes a two-track approach. First, a lot of sensory information is discarded before it gets to the brain. Second, the information destined for LTM is compressed. A lot of the sensory information we receive is redundant, and can be recoded in a more efficient format. Computers regularly use compression techniques to recode large files, particularly ones containing visual images and sounds. Some of the best-known computer compression techniques save memory by throwing away some information from the original data file. These are called "lossy" techniques because some details are lost. The discarded information has to be selected carefully, of course, to make sure that the reconstructed images or sounds still convey most of the essential information.[23]

Human memory is "lossy," too. Suppose, while at your local supermarket, you encounter an old friend you haven't seen in ten years. A week later, it's unlikely that you'll be able to recall all the details of that meeting. You might remember being in the produce section when you bumped into your friend, but you're likely to have forgotten what items were in your basket at the time, what the price of Granny Smith apples was that day, and whether or not your friend was wearing sandals. Those details were probably available at the time of your encounter, and you may have remembered them long enough to recount the events to your spouse while standing in the checkout line. But a week later, those trivial details are missing from your memory of the encounter.

But there are ways to reconstruct some of this seemingly lost information.[24] Events occur within a context, and that context provides background clues that help us remember the details of those events. Perhaps when grocery shopping you always start in the meat department, in which case you "remember" that you had chicken in your basket. Knowing that it was damp and chilly outside that day, you "remember" that your friend had on shoes, not sandals. This knowledge, although not really part of your actual memory of the encounter, helps you reconstruct the meeting, filling in the details. Sometimes this process of memory reconstruction occurs without your being aware that it's happening.

Memories That Play Us False

Simon Van Gelder used to be the assistant director of the penal colony on Tantalus V (TOS-2, "Dagger of the Mind"). But when he escapes from the colony and stows away aboard the *Enterprise*, he is violent and incoherent. Captain Kirk and Dr. McCoy question Van Gelder, but he has a hard time explaining what has happened to him.

> **VAN GELDER:** I was a graduate of . . . I was assistant to . . . I knew . . . I knew, but . . . They've erased . . .
>
> **KIRK:** Erased?
>
> **VAN GELDER:** Edited. Adjusted. Subverted me. [shouting] But I won't forget. I won't forget.

Van Gelder's problem is that much of his memory has, as he puts it, been "erased, edited, adjusted" by a neural neutralizer, a device used in the penal colony. In a way, this erasure of already formed memories, including-long term memories, puts Van Gelder —and other victims of the neural neutralizer—in the same predicament as *Voyager's* holodoctor: they suffer retrograde amnesia.

But the neural neutralizer does something more insidious than merely erasing old memories: it can replace those memories with new ones, as Captain Kirk learns firsthand. Dr. Helen Noel,

Kirk's attractive science advisor on the Tantalus mission, uses the neutralizer to erase Kirk's memory of their first meeting at the *Enterprise's* Christmas party, at which the two of them chatted briefly about the stars. In place of that memory, Dr. Noel plants in Kirk's mind a new memory more to her liking. In Kirk's newly minted memory of their meeting, he fell madly in love with Dr. Noel, literally swept her off her feet, and carried her back to his cabin for a romantic interlude. Kirk vividly remembers things that did not happen. He has "false memories."

You don't need a neural neutralizer to produce false memories, particularly when those memories involve autobiographical events. These erroneous memories come with all the autobiographical trappings of valid ones: you remember where you were during the event and how you felt about what was happening. In short, your memory of the event has become a real part of your remembered past.[25] False memories are very different from run-of-the-mill errors in memory, when you forget a phone number or a person's name. False memory is not a loss of memory. It's a wholesale rewriting of an autobiographical episode.

False memories are not attributable to a design flaw in the human brain. On the contrary, they arise as natural by-products of memory's reliance on reconstruction.[26] As we saw above, one way in which your brain handles the relentless torrent of incoming sensory information is to reduce that information to manageable proportions: it stores only the broad outlines of an episode and fills in the details at the time of recall. This interpolation process is guided by your preexisting knowledge and beliefs as well as by your current needs and expectations. The result? Things happening before and after the episode can influence your reconstruction of the event.

A particularly revealing variety of false memory is confabulation, often described oxymoronically as "honest lying." In the aftermath of damage to the frontal lobes of the brain, a confabulating patient makes statements that are obviously untrue and even self-contradicting. A confabulating patient doesn't intend to lie; in fact, he usually doesn't recognize the untruth of his statements, no matter how clear the evidence. To compound the problem, confabulation can spill over from the patient's erroneous memories into his actions. Take the case of a sixty-one-year-old hospitalized patient, H. W.[27] Over and over again, H. W. insisted that he was *not* in a hospital, but in his office at work. Each evening when his day's "work" was finished, H. W. packed up and tried to head home. Fortunately, the staff redirected him, keeping him safely in the hospital.

Studies of patients like H. W. show that confabulations are not random creations. Instead, these illusory memories are built around the patient's real past experiences. Like false memories, honest lying draws on background information and uses the normal mechanisms of memory to modify and recombine various parts of that background. Memories associated with very different time periods or with very different locations can be uniquely— and erroneously—assembled into novel recollections. In H. W.'s case, the official-looking hospital building triggered the retrieval of a memory of a different location (being at work).

The rewriting of memory is not limited to people with brain damage or to people who have been worked over with a neural neutralizer. Elizabeth Loftus has provided some good demonstrations of the ease with which confabulation can occur in normal, healthy people. After watching films of automobile accidents, viewers were questioned about what they'd seen. Subtle changes in the wording of the questions produced dramatic changes in the way the accidents were remembered. For example, "How fast were the cars going when they smashed into each other?" produced a 25 percent increase in reported speed compared with "How fast were the cars going when they hit each other?" A single word made a world of difference in what people remembered. Imagine what effect these subtle memory "hints" could have in courtroom interrogations, or in counseling sessions aimed at "recovering" repressed memories.[28]

Loftus's volunteers and Captain Kirk have something in common. For both, memories of actual events (an auto accident; the Enterprise's Christmas party) were modified by subsequent information (Loftus's question; Dr. Helen Noel's suggestion). A kernel of truth—an actual event—was colored by an imaginary event. Sometimes false memories haven't even a kernel of truth to work with: they are made up of whole cloth. The authors of this book have a friend whose memory, like ours, occasionally goes beyond reality—he "remembers" doing something that he never did. Recently, he clearly remembered sending an e-mail message to one of us. In his mind's eye, he saw himself sitting at the computer composing and sending the message. He was miffed when he received no response, but the reason for our failure to respond was simple: he never actually wrote or sent anything. He confused his own vivid, internal imaginings with real, external events.

Mentally composing the message created the clear but mistaken conviction that the act had been performed.

There may be good reasons for this wholesale fabrication of events. Imagining an event in your mind's eye activates some of the very same brain areas that are engaged during actual visual perception of objects and events.[29] Because vision and visual imagery share some of the same brain circuits, a vivid image of some event might later be confused with the real thing. Memory researchers describe these kinds of confusions as "failures in source monitoring."[30] You attribute a memory to the wrong source. It's as if you were recounting a story about a green, three-headed alien baby—you attribute the story to an article in *The New York Times*, when in reality you saw it on a sensationalistic late-night cable television program. Sources do make a difference.

False memories can also be synthesized through the power of suggestion—or, more technically speaking, through the power of association. When you see an object or read a word, other items associated with it come into your mind. By exploiting this process of association, it is possible to create false memories to order.[31] Here's an exercise that demonstrates this point.[32]

The box below contains two lists of words. Read the words silently to yourself, one word at a time. Then grab a pencil and complete the test on the facing page.

---

bed, rest, awake, tired, dream, wake, snooze, blanket, doze, slumber, snore, nap, peace, yawn, drowsy

hot, snow, warm, winter, ice, wet, frigid, chilly, heat, weather, freeze, air, shiver, Arctic, frost

---

At this point you should have read through the words in the box and used the test to assess your recognition memory for them. If you're like the majority of people, you made some interesting mistakes. About half the people tested indicated that the word "sleep" was among the words listed in the box. Looking back at those words, though, you won't find "sleep." Similarly, many people claim to have read the word "cold," but it's not there either. Whatever false recognitions *you* made (if any) were probably made with confidence—you clearly remembered having seen certain words, even though they weren't actually listed.

➤ ➤ ➤   TEST   ➤ ➤ ➤

Look at the words listed below and check the ones that you believe appeared in either of the two lists on the opposite page. Once you've finished, return to that page.

___ snore	___ frost
___ heat	___ guide
___ dance	___ snow
___ tired	___ warm
___ dream	___ winter
___ balloon	___ flower
___ snooze	___ wet
___ blanket	___ belt
___ breakfast	___ chilly
___ slumber	___ family
___ fever	___ cold
___ sleep	___ freeze
___ lawn	___ air
___ frigid	___ shave
___ shiver	___ awake

➤ ➤ ➤

False memories of autobiographical details are almost as easy to create as false memories of words. In one recent study,[33] college student volunteers were repeatedly asked about some autobiographical event that had never really happened. A volunteer might be questioned, for example, about her birthday party at which anchovy pizza and pony rides had been featured. Over time, with repeated questioning about this fictitious event, about one-quarter of the volunteers developed false memories of the event, in effect rewriting their own autobiographies.

Even a towering intellect doesn't immunize one against false memories. Jean Piaget, the distinguished Swiss psychologist, reported that for years he'd had a clear but totally false autobiographical memory. Like other members of his family, Piaget believed that when he was very young, his nursemaid heroically thwarted a kidnapping attempt. (For this act of heroism, she was well rewarded by the family.) For many years, Piaget lived with a clear mental image of the attempted kidnapping. When Piaget was in middle age, the nursemaid confessed that she'd made up the whole thing. None of it ever happened. Presumably Piaget's mental image and "memory" of this nonincident were produced by the family's frequent retelling of the story, much as frequent questioning of volunteers can induce memories of events that did not happen.

Physiology of False Memories

Star Trek presents several instances in which false memories are created by outside forces. We've already mentioned Captain Kirk's false memory of falling in love with Dr. Helen Noel (TOS-1, "Dagger of the Mind"). The producer of this false memory was Dr. Noel herself, and she knew exactly what she was doing. Deanna Troi also experiences memories of events that never happened, including visions of being sexually assaulted (TNG-5, "Violations"). These unpleasant false memories, too, were deliberately created, this time by the malice of an alien telepath from the Ullian species. *Voyager's* security chief Tuvok is haunted by terrifying childhood memories involving the death of a young child. These turn out to be false memories resulting from a virus that is consuming peptides in his brain (VOY-2, "Flashback"). And B'Elanna Torres's terrifying autobiographical dreams come from memories implanted in her by a desperate empath who was temporarily on board *Voyager* (VOY-3, "Remember").

In each of these twenty-fourth-century cases, a medical tricorder or other sophisticated device spots abnormal activity in the victim's brain. This abnormality is a telltale sign that the unwanted memory is not homegrown, but has been implanted. The Emergency Medical Hologram informs B'Elanna that "cortical theta wave readings indicate that you were experiencing memories, specifically, *implanted* memories."[34] Diagnosing Troi's memory problem, Dr. Crusher notes unusual activity in the patient's thalamus, which Crusher describes as "the memory center."[35]

Today, there's only one surefire way to distinguish real from false memories. Keep in mind that we're talking about memories that are confused with the truth, not about bald-faced lies. False memories are genuinely believed by their rememberers, including honest, upstanding rememberers. You can't base your judgment solely on the memory's detail. A false memory can be as elaborate as the real McCoy. The only foolproof way to tell a true recollection from a false one is to track down evidence that corroborates or refutes the recollection. This evidence could come from telephone logs, travel records, or from other individuals present at the time of the alleged, remembered incident.

Things may change in the future, however. Neuroscientists have recently begun to develop rudimentary alternatives, exploiting what *Star Trek's* twenty-fourth-century scientists know—that true and false memories seem to generate different neural signatures in the brain. Remember those two lists of words that fooled you into remembering words that weren't really there? Daniel Schacter and colleagues[36] employed a similar test, taking PET scans of the brain while human volunteers judged whether or not visually presented words were among those previously read to them. Their goal was to compare levels of activity in various regions of the brain when the volunteers made correct recognitions with those when they made false ones. Here we'll emphasize three results: a brain region that was active during both correct and false recognition, a region that was active only during correct recognition, and a region that was active only during false recognition.

With correct and false recognitions alike, blood flow increased in the left medial temporal region, a part of the brain that is essential for declarative memory. When you recall what you had for lunch yesterday, or what words you read in those lists—whether your recall is correct or not—you're utilizing brain circuits that include this region. This is the part of the brain Dr. Crusher stimulates to recover lost memories in the amnesic crew members of the *Enterprise*-D (TNG-5, "Conundrum"). And when that region of the brain is damaged by stroke or injury, declarative memories cannot be recalled.

When Schacter's volunteers *correctly* recognized a word they had previously heard, their brains showed heightened activity in more lateral regions of the temporal lobe. These are the brain regions that process and analyze information about the sounds we hear. Thus visual presentation of a previously heard word seems to reawaken our auditory sensory memory of the word, something that new words fail to do.

Finally, a third brain area was active mainly when *false* recognitions occurred. This region, the orbitofrontal cortex, is located just above the eye sockets (*orbit* is another name for "eye socket"). Neurons here become active when you must exert effort to remember something. When remembering is effortless and easy, the orbitofrontal cortex remains quietly inactive. Perhaps the volunteers, though confident about their incorrect memories, had to work harder to "retrieve" those memories. We do know that when false recognitions are made, people take a little longer to reach a decision than when they make correct recognitions.[37]

So contemporary neural imaging techniques represent a promising start toward distinguishing true memories from false ones. But we're a long way from Dr. Crusher's or the Emergency Medical Hologram's unerring ability to tell a false memory from the real thing.

Escape from Time's Prison

Memory does more than simply recording facts, events, and learned skills. Our memories immortalize important but fleeting sensory experiences, allowing us to think about and react to objects and events that are no longer present. They can create temporal anomalies reminiscent of those that show up so often in *Star Trek* (e.g., VOY-3, "Future's End," I). These temporal anomalies allow humanoids to travel backward or forward in time, sometimes, despite Starfleet's Prime Directive, altering the course of history. But you don't need spaceships and warp drives to create a temporal inversion. You can create your own, and your vehicle for time travel is your memory.

Memories lift us out of time's seemingly inexorable forward march, and they do this in two complementary ways. First, memories allow us to revisit events that have already occurred, making it possible to reexamine the circumstances surrounding those events and, perhaps, alter our interpretations of them. Second, our memories of past experiences enable us to project ourselves mentally into the future, anticipating events that have yet to occur and evaluating the likely consequences of various scenarios—what if's.

The key to successful mental time travel is predictability, an often overlooked but essential feature of our own planet, and of every other world encountered in *Star Trek*. Because our world is generally predictable, a record of the past provides a road map to the present and future. Memory is fallible, and it certainly is idiosyncratic. But it is *the* record of our life's journey. So long as we can remember how we've handled our pasts, we have decent odds of being able to handle our futures.

Disordered Brains, Disordered Minds

Pretty arrogant, the human brain. It fancies itself the master of the universe, but when you get right down to it, it's a lightweight, weighing in at a mere 3 pounds (soaking wet). In the flesh, it's not much to look at. It's constructed out of low-tech materials, cobbled together in a shape that only a post-modernist could love.

And this self-styled master of the universe is quite fragile to boot. When a hard object—say, a windshield, or someone's fist—hits a human head, the brain gets bounced around inside the skull's bony plates. If the pounding is hard enough, or goes on long enough, brain cells can be damaged, even killed. And then there's the brain's vulnerability to disease. If a circulatory problem, such as a stroke, pinches off its blood supply, the brain is deprived of the oxygen and energy it needs to survive. In just a few minutes, oxygen-starved neurons are dying by the thousands.

Once neurons die, due to blows to the head, or starvation, or even old age, they're lost forever—the brain is unable to manufacture new ones to make up for the loss. As Julian Bashir, Deep Space Nine's chief medical officer, puts it, "The brain is the spark of life that can't be replicated" (DS9-3, "Life Support").

And when the brain is not being choked by vascular disease or pummeled with blows, it has to deal with invasions of foreign chemicals that the brain's owner eats, drinks, injects, smokes, or snorts. These chemicals—drugs, we call them—can poison brain cells. Even when they don't kill brain cells outright, they alter the normal operation of those cells, often not for the better.

Brain disorders, whether they come from disease, injury, or substance abuse, debilitate the victim, physically and mentally. Boxer Muhammad Ali's brain was damaged by repeated blows to his head. The battering and death of a small group of neurons deep in his brain turned a graceful and boisterous world heavyweight champion into a palsied and subdued shell of his former self.[1]

Human lives and families can be destroyed by brain disorders—they're conditions that no one should have to experience. When such tragedy strikes, it's impossible to avoid a simple truth: disordered brains produce disordered minds and disordered bodies. This is just as true in the future we see when we watch *Star Trek* as it is today. But here's a tiny silver lining to the cloud of brain disorders: these neurological tragedies offer a unique window into the intimate relationship between the mind and the brain. By showing us what happens when things go wrong, these accidents of nature reveal how the brain works when things are going right.

When the Brain Speaks, the Mind Listens

Miles O'Brien is struggling to recover from a harrowing experience (DS9-4, "Hard Time"). After being wrongly convicted of espionage, O'Brien ended up doing twenty years in prison. But this was no ordinary prison term. His jailers, a race of aliens called the Argrathi, devised a punishment that was fiendishly efficient. The Argrathi implanted in O'Brien's brain false memories of twenty years spent in prison. The memories forced O'Brien through two decades of agonizing punishment in what really amounted to just a few hours. But for O'Brien, the experience was protracted and quite real, complete with cellmates, brutal guards, and near-starvation— the works. After he gets out of this mental prison, the last two decades of his life are a painful memory.

O'Brien has returned to Deep Space Nine, his home station, and is struggling to rebuild his life. But his simulated punishment has left some very real emotional scars that cut deeply into his psyche. He fights with his wife; he snarls at friends; he can't concentrate on his job. O'Brien is at suicide's edge.

To ease the pain, O'Brien goes into Quark's bar, orders a synthale (a drink of nonintoxicating synthetic alcohol), and settles alone at a table. Suddenly there appears beside him a man—Ee'Char—who was his cellmate in prison. O'Brien knows that Ee'Char died during their captivity. In fact, he blames that death on his own self-ish refusal to share a morsel of food with his cellmate. O'Brien is naturally upset to find a dead man, particularly *this* dead man, sitting next to him in the bar.

EE'CHAR: Miles.

O'BRIEN: Ee'char, what are *you* doing here?

EE'CHAR: I've never really been gone, have I?

O'BRIEN: You're not real. You're just in my head.

EE'CHAR: That's all I ever was. But I'm real to *you*. And that's all that matters.

Other people in the bar watch, perplexed, as O'Brien carries on a conversation with an empty chair. However real Ee'Char is to O'Brien, this former cellmate is a figment of his tormented brain.

O'Brien is experiencing a hallucination, a visual and auditory experience fabricated entirely by his brain with no help from the external world.[2] And it is no comfort to him to know that the experience is a hoax his mind is playing on him. This hallucination is as real and compelling as the genuine article would be.

Hallucinations are different from more common sensory errors in which you merely mistake one sound or sight for another. Have you ever carried on a phone conversation thinking you were talking with one person, only to discover after a few moments that it's really someone else? Mistakes like this come from errors in interpretation. Your senses provide your brain with appropriate input, fresh from the external world, but that input gets misinterpreted. Something very similar happens in the case of illusions, in which an object's appearance is distorted by the context in which it appears (see Chapter 5). Hallucinations are different from both sensory errors and illusions. Here, input from the external world is missing altogether—the whole experience is created inside your brain, with no help from the environment.

Hallucinations are also different from auditory or visual images that you generate intentionally. If asked to, you could probably conjure up a vivid visual image of the house you grew up in, seeing it with your mind's eye.[3] If the image is vivid enough, you may be able to stand on your old front steps and see the door-bell button that was to the right of the door (or was it the left?). This visual image resembles a hallucination because it occurs in the absence of external input. But these images are different. You never confuse them with the real thing. It's very unlikely that you'll reach out to ring that imaginary doorbell.

The resemblances between self-produced images and externally produced perceptions are no accident. When you intentionally generate a visual image of some scene, you turn on some of the very same brain centers that would be activated if you actually looked at that scene.[4] This arrangement lets imagination and sensory reality time-share parts of the brain's neural machinery. An efficient arrangement—but one with a cost.

The brain's intertwining of imagination and perception can cause confusion between the two. In one demonstration of such confusion, adults in a completely dark room were asked to visualize in their mind's eye a certain object, such as a banana.[5] During these imaginings, unbeknownst to the participants, barely visible, faint, blurry drawings were projected onto a screen directly in front of them. Typically, these projected images coincided with the objects they had been asked to imagine—a banana was shown while the person was imagining a banana.

Virtually none of the people tested had a clue that they were actually *seeing* things, not just imagining them. One person was surprised that the banana he imagined was standing up, on its end. The banana's upright posture struck him as odd, since he was sure that the banana he had tried to imagine was lying on its side. Of course, there was a simple explanation for his error: what he thought was the product of his imagination was actually a real picture of an upright banana. Other people unwittingly combined elements from real and imagined objects, such as a leaf (the real object) laced with red veins (imagination).

Even Starfleet's best, most seasoned officers sometimes confuse the imaginary and the real. Geordi LaForge, chief engineer aboard the *Enterprise*-D, almost dies trying to obey the orders of an entity he erroneously imagines is his mother (TNG-7, "Interface"). Alien life-forms are trapped inside a damaged starship, the U.S.S. *Hera*, and their only means of escape is to get someone to pilot the ship back into the atmosphere of their home planet. The *Hera's* crew members are dead, killed accidentally by the aliens. When LaForge uses remote neural control to operate a probe on the *Hera*, the aliens manage to infiltrate his brain and create a hallucination in his mind. He is sure that he can see and speak with his mother (who, in fact, is lost in space and presumed dead).

In everyday life we rarely confuse the imagined with the real. If you sing "Happy Birthday" to yourself, you do not believe that what you hear is being sung by someone else. (Hum a few bars and see.) Even if you sing it silently, so that you hear it only in your mind's ear, you know that the song is your own doing. The brain keeps real objects and real events at arm's length from imagined ones. But it's not entirely obvious how the brain manages to do this. Because some of the same machinery is used in processing both kinds of input, you'd think there would be ample opportunity for confusion. To understand how the brain avoids confusion, we can look at disordered brains, in which confusion is more common. The particular disordered brains we will consider are the brains of schizophrenics.

Ripping the Web of Thought

Schizophrenia. The name, which means "fragmented mind," was chosen because this disease seems to shred the mind's tightly woven web of thought and feeling.[6] Schizophrenia makes thoughts run every which way but straight. It also brings on delusions and inappropriate emotional responses, such as laughing while you're at a funeral.

Hallucinations are another common symptom of schizophrenia. These hallucinations most commonly involve sounds, particularly voices. Sometimes the voices talk to the hallucinator, giving him advice or orders. Other times, the voices talk *about* the hallucinator. They may seem to be voices that are overheard, gossiping about him or plotting against him.

Imaginary voices can be dangerous, as Lieutenant Commander LaForge discovers when he obeys the orders of his imagined mother. An infamous twentieth-century counterpart was New York City's "Son of Sam" killer. During the 1970s he terrorized the entire city, racking up six random killings and a number of near-misses. When he was finally caught and given a psychiatric exam, the killer was diagnosed as schizophrenic. He confessed to the murders, explaining that he had merely followed orders—the barking of a Labrador retriever owned by his neighbor, Sam.[7]

Let's get one thing straight: The vast majority of schizophrenics are not homicidal maniacs. In fact, not every schizophrenic even has hallucinations. Restricting our discussion to schizophrenics who do hallucinate, it is impossible to explain why one person "hears" an order to kill, while another "hears" something harmless. A look at where hallucinations come from provides a first, small step in understanding why they vary so.

We've already mentioned one of the brain's time-sharing arrangements, in which certain neural structures do double duty, serving the needs of both visual perception and visual imagination. When schizophrenics "hear" things that aren't there, the auditory regions in their brains are making use of a time-sharing arrangement of their own. But this time-sharing doesn't cause the hallucinations. Instead, it makes it hard to distinguish externally produced sounds from internally generated ones.

Consider the following study.[8] A volunteer lies on his back, eyes closed. While he performs various tasks, his brain activity is measured using PET, a neural imaging technique that shows the levels of neural activation in various areas of the brain (described in Chapter 2). If the volunteer is a schizophrenic who suffers hallucinations, he presses a button whenever he hears voices. By keeping track of when the button is pressed and when it is not, researchers can separate brain activity that occurs during auditory hallucinations from brain activity that occurs in their absence. This technique makes it possible to identify areas of the brain that are active (or inactive) only during hallucinations.

When schizophrenic patients are experiencing hallucinations, their brain activity resembles what you find when normal people engage in inner speech: speech and hearing areas are activated.[9] The obvious conclusion? Auditory hallucinations occur because the schizophrenic hears his own inner speech. But before we close the book on this question, notice that there's something missing from this explanation. Suppose you're running to catch a train. As you silently urge yourself on, your own words of encouragement activate speech—and hearing-related areas in your brain. But you never make the mistake of thinking that someone else is speaking; you know it's you. And that's a key to the correct attribution you make ("I am talking to myself") and the incorrect attribution the hallucinator makes ("Someone else is talking to me").

Particular areas of your brain become active when you talk to yourself, but are silent when someone else talks to you. Essentially, when they are active, these areas generate a neural signature that says "This speech is my own doing." For some unknown reason, the brains of hallucinating schizophrenics don't seem to generate such a signature.[10] As a result, those brains fail to identify neural activity in their language regions as self-produced. In the schizophrenic, the brain's impeccable logic goes something like this:

1. Someone spoke (neural activation of the auditory regions of the brain).

2. My owner did not speak (no neural label showing that the speech was self-produced).

3. Therefore, someone else must have spoken.

The internal consistency of this logic may be flawless, but the conclusion is dead wrong.

"Major, Larks True Pepper"

It's happening to just about everyone on board Deep Space Nine: Jake Sisko, Jadzia Dax, Julian Bashir, Kira Nerys, even Captain Benjamin Sisko. They're all speaking gibberish. Chief Miles O'Brien was the first. He'd been under pressure, having to fix things all over the station: a jammed airlock, malfunctioning replicators, buggy computers. While working with O'Brien in the Operations Center—called "Ops" for short—Kira Nerys notices how stressed he seems, and tries to help (DS9-1, "Babel").

> KIRA: I suppose this isn't a good time to tell you that Number Three Turbolift is broken down again? Joking, Chief.
>
> O'BRIEN: Major, larks true pepper.
>
> KIRA: What!
>
> O'BRIEN: Let birds go further loose neck . . . Shouts easy play.
>
> KIRA: Chief, you're not making any sense.
>
> O'BRIEN: Round the turbin and quick.
>
> KIRA: I . . .
>
> O'BRIEN: Well, close the reverse harbor . . . [Kira shakes her head in disbelief] Ankle. Try. Sound . . . Reset gleaming . . . Dead dinner to burn.

O'Brien can tell by her quizzical look that Major Kira has no clue what he's saying, and her incomprehension makes him so angry that he stomps out of Ops in a huff. Kira intercepts O'Brien and takes him to the infirmary, where Dr. Bashir runs some diagnostic brain scans. O'Brien's visual cortex and auditory cortex are perfectly normal, so they're not the problem. But what is? Sensing

Bashir's perplexity, O'Brien grabs a touchpad and types in the following message:

Flame the dark true
salt way link
complete strike
limits victory
frosted wake
simple hesitation

Looking at the message pad, Kira is dumbfounded, but Dr. Bashir understands: O'Brien is suffering from *aphasia*. His thought processes are normal, but he can't make himself understood—and he can't understand others.

Bashir's diagnosis is just what you'd get today from any neuropsychologist or neurologist. Aphasia is a neurological disorder described in the nineteenth century by the French neurologist Paul Broca[11] after one of his patients lost the ability to speak. When this patient conveniently died—just a few days after Broca had first examined him—Broca performed an autopsy, which pinpointed damage to a particular spot in his patient's left frontal lobe as the cause of his loss of speech. In honor of Broca's discovery, this part of the brain is now known as *Broca's area* (see illustration on facing page).

Not that aphasia is a single entity. The disorder actually takes several different forms, depending upon the location and extent of the brain damage. This connection between symptomatology and underlying neural cause lets us exploit O'Brien's own strange words to figure out what has happened to him.

O'Brien's speech consists of phrases strung together in normal, sentence-length utterances. The rhythm and articulation of his speech seem normal, too. (You can verify this by reading some of O'Brien's statements aloud.) If you ignore the actual words, the overall sound resembles that of normal English. These observations rule out a diagnosis of *Broca's aphasia* (which is characterized by short, choppy, nongrammatical speech). So we know that the problem does not reside in O'Brien's left frontal lobe.

Three symptoms hold the keys to O'Brien's problem. First, his speech has a normal pace and rhythm, even though what he says is incoherent and meaningless. Second, he cannot understand what other people say, but he can sense the emotional tone of what's being said. For example, he can tell from their tone of voice that

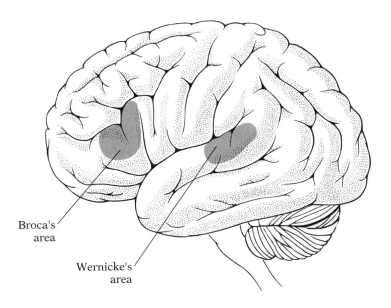

Broca's
area

Wernicke's
area

other crew members are very worried over the spreading condition, but what they're actually saying is beyond him. Third, his written communication is every bit as incomprehensible as his spoken communication.

This trio of symptoms tells us that O'Brien is suffering from *Wernicke's aphasia*. This variety of aphasia results from damage to a part of the brain's temporal lobe known as *Wernicke's area* (see illustration above). In any particular case, the severity of the symptoms depends upon how much tissue has been damaged. As with Broca's aphasia, the brain damage associated with Wernicke's aphasia is nearly always confined to the left hemisphere, at least among people who are right-handed.[12] By the way, patients with Wernicke's aphasia begin by strongly denying that there's anything wrong with them. They blame others for their communication difficulties, just as O'Brien does.

O'Brien's aphasia is unusual because it comes from a viral infection, not from a stroke or a severe blow to the head. According to Dr. Bashir's analysis, the virus has disrupted the neural pathways that link the auditory perception of words with their stored meanings. A virus can actually do this. In the twentieth century, several cases of aphasia have been traced to the herpes simplex virus, which has a nasty habit of attacking the brain's temporal lobes.

The several varieties of aphasia tell us that speaking and understanding language do not depend on a single "language center" in the brain. Instead, these functions emerge from a complex interplay among many different brain areas, each doing its own job and sharing information with the others. Language depends upon delegations of neurons, not just a single chief executive. Normally, the functions of these different brain regions are smoothly orchestrated, allowing us to generate and understand sentences of infinite variety. Some of us can even do these complicated tasks in several different languages, even without the aid of Starfleet's old standby, the universal language translator.

Fears and Panics

Starfleet is very proud of the warriors and heroes who have served aboard its flagship, the U.S.S. *Enterprise*-D. But Lieutenant Endicott Reginald Barclay is neither a warrior nor a hero. He is perpetually nervous and twitchy. He avoids people whenever possible, and when he does muster the courage to venture into the *Enterprise*-D's bar/social hall, he orders warm milk. His quirkiness and constant whining make him as unloved as the vegetable for which he is nicknamed, "Broccoli" (TNG-3, "Hollow Pursuits").

But shyness and a fondness for warm milk are not Barclay's only problems. He is deathly afraid of transporters—which is somewhat understandable. Think about what happens when you use a transporter. First, the device converts your body's matter into energy. At this point, you no longer exist, in any conventional sense of the word. Then your energy is beamed at incredible speed to another location, where it's transformed back into your original, material form. At least that's the way it's supposed to work. Early models of transporters, in the twenty-third century, occasionally had problems, as we saw in Chapter 2. In addition to those problems, transporting sometimes caused a breakdown of the brain's neurochemical molecules. This alteration in brain chemistry produced a highly disturbed mental condition known as transporter psychosis (TNG-6, "Realm of Fear").13

Reg Barclay's fear of transporters, however, is so powerful that it disrupts his Starfleet duties. He desperately searches for ways to avoid the transporter, preferring to travel by shuttlecraft, a much slower means of transportation.14 And he is unable to talk himself out of his fear. He has enough technical knowledge and expertise to know that twenty-fourth-century transporters are at least as safe as shuttlecraft, but that's not enough.

All of us have a little Reg Barclay in us. You can probably think of situations—such as public speaking—or objects—such as snakes—that make you very anxious. And because our world is filled with dangerous objects and threatening situations, your anxiety is not unreasonable. In fact, it reminds you to be careful, to tread cautiously. But when anxiety is excessive, it becomes a persistent, irrational fear called a *phobia*. And when this happens, cautionary anxiety turns into disabling paralysis. Reg Barclay's fear of transporters certainly qualifies as a phobia. It is strong, irrational, and incapacitating.

It is very, very hard to talk yourself out of a phobia, even when you know it's irrational. Garak, the Cardassian tailor and part-time spy aboard Deep Space Nine, can testify to the imperviousness of phobias. Anyone with a sense of anxiety about tightly closed spaces will suffer along with Garak during his episode of claustrophobia. Here's what happens.

The Jem'Hadar and the Vorta, agents of the Dominion, have captured several prisoners: Dr. Bashir, Commander Worf, Klingon General Martok, and Garak (DS9-5, "By Inferno's Light"). The prisoners' only hope of escape is a real long shot. They would have to alter the prison's electrical system, then use the altered system to send an S.O.S. to Deep Space Nine's runabout. The group does have an electronics expert who could do the job. The problem is that their electronics expert is Garak, and he's claustrophobic, scared to death of enclosed spaces (the result of an accident in which he was buried alive). The space in which Garak would have to work is perfect for making a claustrophobic's blood run cold. It's a tiny, tight, hot, dark gap between the steel walls of the Jem'Hadar prison.

Laying his terror aside, Garak crawls between the walls and begins to work. Soon, however, his fear rises to the panic level. Garak tries to talk himself into calming down:

> *I'm sorry! But that's absolutely unacceptable. I'm under enough strain as it is. I can't have you quitting on me. Get ahold of yourself, Garak. After all, you haven't had one of these attacks in years. Yes, this is a tight enclosed space. Yes, there's not a lot of room to move. But a disciplined mind does not allow itself to be sidetracked by niggling psychological disorders like claustrophobia.*

But not even the disciplined mind of a Cardassian can talk itself out of a phobia. Garak slides into a panic, banging his head and arms rhythmically against the walls. The other prisoners must pull him out of his prison inside a prison.

Garak's reaction illustrates phobia's telltale biological signs. First and foremost, his intense anxiety is accompanied by heightened activity in his autonomic nervous system. This is the portion of the nervous system that controls respiration, heart rate, and blood pressure; it also regulates glandular secretions, such as adrenaline, sweat, and saliva. When your physiological arousal is intense, portions of your autonomic nervous system work overtime, producing arousal's familiar reactions. But for some individuals—Garak included—these autonomic reactions get out of hand, producing what we call a panic attack.[15]

Because the autonomic nervous system operates more or less automatically, outside of conscious awareness, there's little we can do to mentally snuff out the spreading flames of anxiety. The only quick fix is to remove yourself from the frightening situation. This relieves your anxiety; you feel better. And, of course, the terrifying intensity of your anxiety reaction gives you a powerful incentive to avoid the situation in the future. And this avoidance is precisely the recipe for the creation and maintenance of a phobia.

The word "phobia" comes from the name of the Greek god Phobos, who dealt with his enemies in an unusual way: he scared them to death. Thanks to him, the suffix -phobia now signifies a source of intense anxiety, as in hemophobia (fear of blood), scotophobia (fear of darkness), arachnophobia (fear of spiders), agoraphobia (fear of being in open spaces), and Garak's own favorite, claustrophobia. Acrophobia, fear of heights, is one of the more common phobias. You know that the observation deck atop the Empire State Building is perfectly safe. It's open only when wind conditions permit, and it's surrounded by a sturdy, high fence. Given New York City drivers, you're much more likely to die crossing 34th Street eighty-six floors below than by falling off this deck. The rational aspect of your mind knows all this, and you rehearse these facts as you leave the elevator and head out onto the observation deck, determined to look down. But then something strange happens: you just can't bring yourself to face down that fear. You've turned into Reg Barclay. But take heart. As Martok, the distinguished Klingon general, says, "There is no greater enemy than one's own fears" (DS9-5, "By Inferno's Light").

Anxiety—even intense anxiety—about some object or situation doesn't qualify as a phobia. That term applies only if the anxiety is persistent and interferes with your daily life. Anywhere from 5 to 10 percent of the American population has at least one phobia. Many people have one particular phobia, which most of us have experienced in a milder form. Because it's so common—and so often overlooked—this phobia deserves special mention.

Making Friends with a Potted Plant

BARCLAY: You . . . you don't know what a struggle this has been for me, Commander.

LAFORGE: Well, I'd like to help if I can.

BARCLAY: Being afraid all the time—of forgetting somebody's name, not . . . not knowing what to do with your hands. I mean I mean I am the guy who writes down things to remember to say when there's a party. And then when he finally gets there he ends up alone, in the corner . . . trying to look comfortable examining a potted plant.

LAFORGE: You're just shy, Barclay.

BARCLAY: Just shy? Sounds like nothing serious, doesn't it? You can't know.

(TNG-3, "Hollow Pursuits")

At one time or another in their lives, many people experience a version of Lieutenant Barclay's phobia: social phobia.[16] Contrary to what its name implies, social phobia is not merely a fear of other people, or even groups of people. It's much more specific and more interesting than that. Social phobia is a persistent, deep dread of situations in which you believe others will be scrutinizing you. A social phobic dreads doing or saying something that might be humiliating or embarrassing. So he avoids situations where there's any chance he could be the center of attention.

Acrophobia, for example, has little effect on the daily life of a modern urban dweller, but social phobia is a real show-stopper. It inhibits all sorts of social activities, from eating in restaurants to using public rest rooms. Of course, you can plan your daily activities to avoid public rest rooms. But social phobics suffer more than inconvenience. Their career choices are narrowed, and their social lives are stunted. Compared with their peers, social phobics have lower average incomes, are less likely to complete a college education, and are more likely to be unemployed.[17]

We can't say what starts it, but social phobia can plunge its victims into an unending downward spiral. The spiral begins when they turn to drugs such as alcohol to take the edge off their fears. By medicating themselves, they're able to perform in public. But the drugs increase the odds that the phobics will actually do something inappropriate or plain stupid, and get (deserved) criticism for it. That criticism, of course, makes them even more fearful of future humiliations. So they take more drugs, and the spiral continues—downward.

In North America and Western Europe, people with social phobia are afraid that they'll embarrass themselves in public. This fear focuses on consequences for the individual.[18] Other cultures in Africa and Asia emphasize not the individual, but the group[19]—what the Borg would call "the collective." This emphasis on the collective produces an intriguing variant on social phobia, called Taijin-Kyofusho, which in Japanese means "against other people" (*taijin*), "to be afraid" (*kyofu*), "disorder" (*sho*). Essentially unknown in Western psychiatry, Taijin-Kyofusho is characterized by a fear that inappropriate behavior or appearance will offend others.[20] Like a Western social phobic, a person with Taijin-Kyofusho tends to avoid social situations, but does so because of worry about insulting other people, not about self-embarrassment.[21]

Getting Hold of Yourself Is Hard to Do

When Garak has his claustrophobic reaction in that tiny, dark, hot, confining space, he tries to talk himself out of panicking. But he fails to get hold of himself, and that failure makes things worse. Inability to control oneself is one of phobia's most humiliating and disturbing side effects. But fear of losing control can be more than a mere annoyance or embarrassment.

For victims of phobia's most dramatic and frightening relative, the panic attack, fear of losing control may be the heart of the problem.[22] With no warning, the victim of a panic attack is suddenly overwhelmed by anxiety, a feeling that something terrible is going to happen. The symptoms include sweating, chest pains, numbness, nausea, gasping for breath, and a runaway heartbeat. Often the helpless victim of a panic attack is sure he's going to die of a heart attack. In fact, this certainty that doom is on the doorstep is one of the disorder's defining characteristics. A panic attack can be as short as a few minutes or last for many hours.

When your body's physiological responses are subtle or ambiguous, how you interpret those responses takes on great significance. Tightness in your chest might be a sign of heart trouble; then again, it might be a sign that your muscles are sore from exercise. Ditto for sudden shortness of breath. Folks with panic disorder tend to interpret these ambiguous responses as threatening rather than benign.[23] This doom-and-gloom interpretation creates what students of panic disorder dub "fear of fear."

To understand what goes on in the brain when a person is experiencing extreme anxiety, researchers have turned to brain imaging techniques. One group of researchers took PET scans of volunteers who were normal except for their profound fear of spiders—arachnophobia.[24] In one test session, the volunteers watched a "neutral" videotape of people strolling in a park. In another session, they watched a tape of live spiders crawling over a person's hand and arm. Compared with the neutral tape, the spider video produced substantial levels of physiological arousal, including an increased heart rate. Clearly, the spiders made the volunteers very anxious, which also showed up in their numerical ratings of their own anxiety levels.

The volunteers' brains reacted differently, too. In fact, the spider tape evoked two distinct patterns of neural activity. First, it increased neural activity in the visual association areas, the brain regions that process visual information about objects. Heightened activity in these brain areas makes sense: phobic individuals probably become visually extra-attentive when they see spiders. The second pattern was less straightforward: watching the spider video lowered neural activity in the hippocampus and other limbic system structures. These regions of the brain are involved in the storage and retrieval of declarative memories (see Chapter 6). This lowered activity might reflect the volunteers' efforts to damp down awareness of their past anxiety reactions to the phobic stimulus.[25] We would bet that centuries from now, neural activity in the memory-related regions of Garak's brain will decrease when that claustrophobic Cardassian talks himself into entering that confined, dark, hot space between the walls of the Jem'Hadar prison.

This Mental Hospital Drives Me Crazy

I may be surrounded by insanity, but I am not insane.

William Riker (TNG-6, "Frame of Mind")

Commander Will Riker wakes up in a mental institution. He is surrounded by unfamiliar, very unfriendly faces. One of these faces tells him that his entire previous, "real" life has been a delusion fabricated by his mind to protect himself from the horrible truth that he murdered and mutilated someone (TNG-6, "Frame of Mind"). Riker tries to convince the physician in charge of the facility—a humanoid named Dr. Syrus—that there's been a misunderstanding. After all, he knows that he is a Starfleet officer, not a crazed murderer. But Dr. Syrus turns Riker's "belief" against him, explaining that his denial simply confirms the seriousness of his mental condition. His denial is perfect proof that his mind is desperate to hide the dark truth from itself.

Of course, Riker really *is* a Starfleet officer, and he's never murdered anyone (although he has killed in the heat of battle). But in his altered environment, Riker begins to experience trouble distinguishing reality from hallucination. At times he actually accepts Dr. Syrus's explanation that his Starfleet career is a delusion. A startling turn of events, when you consider that we're talking about William Riker, a highly trained, self-confident officer—one of Starfleet's finest.

But Riker is no more gullible than the next person. Inside a mental hospital, or in normal everyday life, how you interpret your thoughts and feelings depends on the context in which they occur. If Dr. Syrus kidnapped you and threw you into a situation like Riker's, you too would begin to doubt your own sanity. Think about it: you find yourself in a mental hospital, and your mental disorder is certified by competent, knowledgeable professionals. Riker's captors paint a very realistic portrait of him as a crazed murderer. The mental institution and its personnel manipulate him into accepting his assigned role of "patient." They escort him to the dining quarters for his meals; his dress and appearance are disheveled; in the commons area, other patients engage in mindless activities such as arranging blocks on a table and playing with clay. One of the simulated patients even holds a spoon and whispers into it, confiding to Riker that the spoon is actually a communications device. (No, it's just a spoon.)

Dr. Syrus created the simulated mental institution specifically to trick Riker into giving up some key military secrets. But suppose that Riker—or you—were wrongly committed to a real mental institution. Could you convince the staff that a mistake had been made?

In the 1970s, one investigator set out to answer this question.[26] Under his supervision, eight normal, ordinary adults volunteered to masquerade as "pseudo-patients." Each volunteer went to the admission desk of a different psychiatric hospital, complaining of hearing unfamiliar voices uttering words like "hollow" and "empty." These symptoms were totally bogus. The pseudo-patients also gave false names to the hospitals' admission staff, and lied about their vocations and current employment. All of this was done to hide their real identities and their real mental status. Other than these fabrications, they did not alter their descriptions of their past history, and they accurately portrayed their relationships with family members and friends. In short, they tried to behave as they normally would, aside from their fabricated complaint.

Each of the eight pseudo-patients was diagnosed as schizophrenic and admitted to the hospital's psychiatric ward. From this point on, they stopped reporting any symptoms of mental disturbance. Now their mission was to behave in a manner that would get them discharged as quickly as possible. How long would it take the psychiatric personnel to realize that these were sane people in insane places? The average hospital stay for the pseudo-patients was nineteen days. The luckiest managed to get released after only a week; the least lucky took fifty-two days to be "cured."

Why did it take so long for the pseudo-patients to be discharged if they were behaving normally?[27] In abnormal situations, normal behavior gets misinterpreted. As pointed out earlier, our interpretation of behavior depends on the context in which it is observed. In a library, we don't give a moment's notice to a person who spends lots of time reading and making notes. But if a patient on a psychiatric ward does the same thing, it can be construed as compulsive behavior, an indicator of mental disorder. The psychiatric staff made exactly such misinterpretations of the pseudo-patients' normal behavior. For example, patients, including the pseudo-patients, tended to gather outside the hospital lunchroom before it opened. One of the psychiatrists chalked this behavior up to the patients' "oral-acquisitive" nature, meaning that they had urges to put objects in their mouths. In fact, they gathered outside the lunchroom for a much simpler reason: there was little to look forward to on the ward aside from lunch.

Labeling someone "schizophrenic" changes the way others look at that person. And, unfortunately, a label that denotes mental disturbance is not like a Post-it that peels right off when you're done with it. A psychiatric label sticks firmly to a patient. All the pseudo-patients in the hospital study were released, but with a fascinating notation on their hospital records: "Schizophrenia in remission." Apparently, once such labels are attached to a patient, they are hard to remove.

The Ultimate "Cure" for Depression

Garak is depressed and angry (DS9-2, "The Wire"). Once he was an important figure in the Obsidian Order, Cardassia's fabled and feared security agency. Now he's a lowly tailor on a Federation space station overflowing with his former enemies, the Bajorans. We're not sure why he left Cardassia, but Garak is not a happy camper. Fortunately, though, there is something he can do about his misery. Implanted in his brain is a device that triggers the release of endorphins.[28] These morphinelike neurochemicals, which are natural products of the human brain, mute pain and can induce a sense of euphoria.[29]

Garak's depression grows to the point where his implant must run continuously, feeding his brain a steady stream of endorphins. Eventually his brain is thrown into neural shock, which forces Dr. Bashir to inactivate the device. His withdrawal symptoms are agonizing, like those of heroin addicts who go cold turkey.

Garak's predicament is understandable—sometimes, life is not all the fun it might be. Almost everyone has periods of sadness, apprehension, or guilt. It's hard to fall asleep, and we wake up in the middle of the night worrying about things. We lose interest in food and sex; we cannot organize our thoughts or finish a chore. The wear and tear of normal living can easily produce any one or even two of these symptoms. But if all these symptoms arrive together, and if they hang on for months or even years, they may be something more than a response to ordinary wear and tear. They could be symptoms of a debilitating mood disorder called depression.

Mood disorders, including depression, are extremes on a continuum that runs from normal at one end to abnormal at the other. When the symptoms of depression are short-term and situation-specific, they don't send you scurrying for professional help. But when they ruin your daily life, when they threaten your health or safety, when they undermine your social interactions—then it's time to look for help. And getting help is especially urgent if your best option seems to be suicide.

On our planet, suicide—intentionally ending one's own life—is the invention and private property of the human species. And it—or at least the contemplation of it—is common in our species. In fact, a majority of the United States' population has had suicidal thoughts at some point during their lives.[30] Suicide is also "contagious." The incidence of suicide spikes right after a suicide story appears on television or in newspapers.[31] And the more publicity one suicide gets, the greater the number of subsequent suicides.[32]

Star Trek's suicides happen for many of the same reasons that suicides happen today, on Earth.[33] Even the most fearless warrior in Starfleet, Worf, has contemplated suicide several times during his illustrious career. His first bout with suicidal thoughts occurs when fear destroys his self-confidence. This could be a disturbing situation for anyone, but for a Klingon warrior, it's absolutely devastating (TNG-4, "Night Terrors"). Worf considers suicide again when an accident paralyzes him from the waist down, leaving him with no hope of recovery (TNG-5, "Ethics"). Evidently bravery does not grant one immunity to the temptations of suicide.

And neither does omnipotence, even when we're talking about the godlike, extradimensional entities called the Q. One of the Q—"rebel Q," as he's called—does a Socrates on himself (VOY-2, "Deathwish")—he commits suicide the old-fashioned way, by drinking hemlock. His motive? Utter, profound boredom with life. Q has done everything there is to do, and he knows everything there is to know. "For us, the disease is immortality," Q says to explain his death wish.

Among twentieth-century humans, suicide's leading cause is depression. In the United States, depression is implicated in about 70 percent of the approximately 30,000 successful suicides that occur each year. Surveys show that many people believe that depression comes from bad parenting or from personal weakness—only a small minority of people surveyed understand that depression has a biological basis in brain function. Yet this biological basis is real, and it has allowed researchers to create remarkable drugs to fight depression's symptoms. These medications, called antidepressants, work by altering brain chemistry. Although the details vary, all antidepressants work by altering neuronal responses to one or more of the brain's neurotransmitters, the chemicals neurons use to communicate with one another.

The widely prescribed antidepressant Prozac works just this way: it promotes the buildup of the neurotransmitter serotonin within synapses. Prozac turns off a normal enzymatic process that usually sweeps the synapses clean of excess molecules of serotonin. By putting a stop to this waste of perfectly good neurotransmitter molecules, Prozac makes more serotonin available to stimulate neurons that need more.

In addition to medication, depressed people can also benefit from psychological therapy that teaches them positive coping skills. Depressed people typically hold low opinions of themselves. They blame themselves for events and circumstances that are far beyond their control. Therapies that help depressed people take a new approach to their life circumstances can relieve depression's symptoms and reduce the chances of relapse.[34]

The Jem'Hadar Are Dying to Kill You

In the Milky Way, no quadrant is free of substance abuse. In the distant Gamma Quadrant of our galaxy, just as in some places on Earth today, many of the most serious acts of violence are linked to drug addiction. The Jem'Hadar, a genetically engineered race, kill for their daily drugs. They are addicted to an enzyme called ketracel, which is known to its addicts as "white." A tube inserted into each Jem'Hadar's neck delivers a continuous infusion of white.

White is a psychotropic drug—it acts on the brain to alter mood, thoughts, and behavior—and it's highly addictive. Without a steady flow of white, Jem'Hadar experience terrible anxiety (DS9-3, "The Abandoned"; DS9-6, "Rocks and Shoals") and eventually die (DS9-4, "Hippocratic Oath"). Unfortunately for the Jem'Hadar, there is only one source for the drug: the Vorta, agents of the Dominion. In return for their white, the Jem'Hadar do the violent bidding of their ruthless suppliers. Their addiction makes the Jem'Hadar slaves—very, very dangerous slaves.

The Vorta are not the only people in the galaxy that use addiction to enslave others. The Brekkians have only one industry on their planet: the production of felicium, a narcotic substance derived from plants (TNG-1, "Symbiosis"). Centuries before, felicium was used for medicinal purposes, to fight a deadly disease that was ravaging the neighboring planet, Ornara. But the cure turned out to be as bad as the disease. The people of Ornara are hooked on felicium without realizing it: they believe they have to take the drug to keep the disease at bay. The Ornarans don't know that the disease is no longer a threat—it's been eradicated. Felicium—for which the Ornarans pay dearly—is no longer needed for medicinal purposes, just for its psychotropic effects. The Brekkians hide this fact from their drug-dependent neighbors. With all that money to be made, they have no motive to reveal the truth.

The Ornarans got hooked on felicium for the same reason that twentieth-century crack addicts get hooked on cocaine. In both cases, small doses of these chemicals powerfully alter an individual's thoughts and behaviors. Psychotropic drugs mimic, enhance, or block the neurotransmitters that carry chemical messages among the brain's billions of neurons. Whenever you think a thought, that idea is synthesized by neural activity arising from your neurotransmitters. Whenever you experience intense pleasure—a bite of Deanna Troi's favorite food, chocolate, for example—you have your brain's neurotransmitters to thank. Whenever you perform an action—whether it's brushing your teeth or slamming your fist against a wall—your brain uses neurotransmitters to generate motor plans and transmit motor commands to your muscles. In brief, our bodies and minds are energized by our brain chemistry. In a sense, we are all "drug addicts," but the drugs we rely on are those designed by Nature to allow our brains to operate efficiently and adaptively. Sometimes these neurochemical systems go haywire, and when they do, the result is a disorder, such as depression. But for most of us, our brain's own built-in drugstore works just fine. At least, that is, until we introduce chemicals from outside into our bodies.

The same actions that are produced by your brain's own natural neurotransmitters can be triggered artificially by chemicals that you ingest, inhale, or inject into your body. These chemicals from outside the brain—drugs—are surrogates that can turn on the brain's neurochemical machinery. They have their own keys to the neural ignition system. And activating the brain's machinery in this unnatural way produces exactly the reactions—thoughts, pleasurable feelings, behaviors—that the machinery normally produces. A brain on drugs isn't doing anything it wasn't designed to do. The problem is that it's doing it when it shouldn't be.

Drugs trick the brain into action in one of two ways, either by masquerading as normal neurotransmitters or by blocking neurotransmitters' access to their targets. Drugs that mimic the action of a neuron's normal neurotransmitter are called *agonists*. These chemical mimics either excite neurons or inhibit them, depending on the type of neurotransmitter they're impersonating.

Consider, for example, the dopamine pathways discussed in Chapter 2, in which neural activity creates a strong sense of pleasure.[35] Cocaine, nicotine, and amphetamines are all agonists that create a sense of pleasure by activating receptors on dopamine-sensitive neurons in these pathways. (These are the same pathways that are stimulated by Garak's neural implant.) These pleasure-producing drugs become addictive because of the enjoyable feelings they create. The addiction is maintained, at least in part, because the drug's absence provokes severe withdrawal symptoms, in which pleasure is replaced by a punishing nightmare.[36] Judging from its behavioral effects, felicium—the drug to which the Ornarans are unknowingly addicted—certainly belongs in this category.

Other drugs operate by blocking access to receptors on the neurons. By hanging up a chemical "OCCUPIED" sign, these blockers make it difficult or impossible for the appropriate neurotransmitters to attach themselves to the blocked receptors. Drugs that operate in this fashion are called *antagonists*, and their effect is comparable to that of sticking a wad of gum into a lock—the key (the neurotransmitter molecules) no longer fits into the keyhole (the neurons' receptors).

One antagonist familiar to nearly everyone is caffeine, a drug widely used both in Western cultures today and in Starfleet. Caffeine inactivates a mechanism that normally limits the level of activity within certain brain cells. Liberated by caffeine, these cells become hyperresponsive.[37] The ultimate consequences, for which millions of people pay billions of dollars, are increased alertness and improved mood.

The Twentieth Century's Favorite Drug

Of all the drugs used on Earth, alcohol is probably the most widespread. All by themselves, Americans consume almost 7 billion gallons of alcoholic beverages a year. That is enough liquid to fill 318,182 railroad tank cars, which, linked end to end, would stretch from Maine to Oregon. This flood of alcohol is consumed largely because alcohol, in moderation, reduces anxiety and produces a sense of euphoria, making it easier to socialize.

Imagine that you're cooped up on the *Enterprise*-D for months at a time, alone with a few hundred other people, hurtling through the void of space. Aside from the rare contact with an unknown alien species or the occasional temporal rift, life on a starship, even the *Enterprise*-D, can get pretty monotonous. Your main entertainments are fantasy programs on the holodeck and visits to the ship's lounge, Ten-Forward.

Ten-Forward offers not only a spectacular view of the stars, but also a nice variety of mixed drinks served up by a bartender— Guinan—who's a wonderful listener. For the crew of the *Enterprise*-D, Ten-Forward is a wonderful escape from the routine of work. With all Ten-Forward's temptations, why don't more crew members develop drinking problems?

The answer is simple: The mixed drinks served aboard the *Enterprise*-D are made with the alcohol substitute synthehol. And the same is true of most of the drinks available in Quark's bar on Deep Space Nine. According to the 253rd Rule of the Ferengi Rules of Acquisition, synthehol is the lubricant of choice for a customer's stuck purse.

Not much is known about synthehol, except that it was invented by the Ferengi and that it has an intoxicating effect without alcohol's negative side effects. Evidently, drinks made with synthehol don't taste exactly like their alcohol counterparts. The well-practiced palate of Montgomery Scott, chief engineer aboard the original *Enterprise*, finds synthehol-based scotch vastly inferior to the real thing (TNG-6, "Relics"). And Jean-Luc Picard's brother Robert, who has lived all his life in the wine country of France, is sure that synthehol has ruined Jean-Luc's appreciation of fine wine (TNG-7, "All Good Things . . ."). Whatever its shortcomings, though, synthehol—at least aboard the *Enterprise*-D— seems to have eliminated one of society's most nagging, serious problems: alcohol abuse.[38]

Unlike many other drugs, alcohol does not directly stimulate brain receptors. Instead, it does its work in a cunning, roundabout way. It alters the effect of GABA, one of the brain's most common neurotransmitters. When GABA binds to its receptors on the receiving end of a neuron, it changes the neuron's electrical state, making the neuron less excitable. This effect is described as inhibition. The result? The inhibited nerve cell is now less responsive to whatever stimulation it may receive. Alcohol strengthens GABA's inhibitory power. It makes a given amount of GABA more effective than normal. And because neurons all over the brain contain GABA receptors, alcohol's depressing effect on brain activity is widespread. In addition to reinforcing GABA's effects, alcohol also stimulates the dopamine-sensitive pathways, which probably accounts for the drug's pleasurable effects.

Knowing how alcohol affects the brain, we can speculate about how synthehol might operate. This Ferengi invention produces alcohol's relaxed feelings and sociability without disturbing motor coordination and judgment. Synthehol's effect on the brain is undoubtedly more selective than alcohol's is. Alcohol, as we have seen, inhibits or depresses activity over virtually the entire brain. Synthehol seems to target the limbic system and, perhaps, regions of the frontal lobes—those brain areas that are most sensitive to dopamine—but spares those areas responsible for motor control and decision making.

Our explanation of synthehol's "magic" may be too good to be true. We don't mean to malign the Ferengi inventors of synthehol, but Ferengi ingenuity and Ferengi interest in maximizing profit do suggest one other possible account of how synthehol might work. It may be that synthehol has no expensive, psychoactive ingredient whatever, but merely produces a "placebo" effect in those who consume it. If Starfleet personnel expect that synthehol drinks will relax them and make them more sociable, this expectation itself could be responsible for creating those feelings. This is exactly what happens if people are told they are consuming alcohol, but in fact are given alcohol-free substitutes. These misled individuals can become aggressive, gregarious, or relaxed, depending on their environment.[39] In short, they behave just as if they had actually consumed a few alcoholic drinks. But their motor coordination and judgment remain unimpaired. This description sounds suspiciously like synthehol's effects.

Of course, for synthehol to work successfully as a placebo, its consumers must really believe that the drink reduces anxiety, increases sociability, and relaxes inhibitions. Once a Starfleet crew member learns that it's all in his mind—not in the drink—synthehol could be robbed of its power. So please don't discuss this possible placebo effect with any Starfleet folk you happen to meet.

Brain Disorders' Silver Lining

So what do Garak, O'Brien, Reginald Barclay, and Muhammad Ali have in common? They all have suffered brain disorders, and as a result, have had to live with disordered minds. Garak and O'Brien are fortunate—their neural abnormalities are cured by surgery and by medication. Barclay's condition is chronic, but not hopeless—he is learning how to keep his fears in check. But Muhammad Ali will remain a tragic victim of that tiny cluster of renegade neurons deep within the interior of his brain—late-twentieth-century neuroscience has yet to develop the magic necessary to restore him to health.

When faced with these kinds of tragedies, we nonetheless are reminded of the aphorism offered at the outset of this chapter: Tragedy can lead to insight. By showing us what happens when things go wrong, disorders of the brain reveal how the brain works when things are going right. And when things *are* working properly, the human brain comes pretty close to actually deserving that title, "master of the universe"—or at least "master of Sector 001 of the Milky Way galaxy."

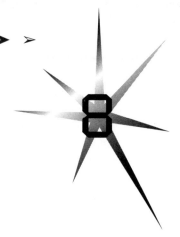

Today's Brain, Tomorrow's Mind

It would be very unlikely for unlikely events not to occur.

J. A. Paulos[1]

Several hundred years separate us from the universe we see in *Star Trek*. But passing centuries and breathtaking technological developments matter little when it comes to basic human psychology. *Star Trek's* human characters think like we do, forget like we do, get angry like we do, and suffer like we do. It's easy to recognize ourselves in them. Hundreds of years from now, far, far from Earth, the behaviors and thought patterns of twentieth-century humans have been "conserved," to borrow a term from genetics.

If biology has anything to say about it, three or four hundred years from now—assuming our species survives until then—our descendants' bodies and brains will be much like ours. Do the arithmetic. New generations appear about every 35 years, so we're only about nine generations removed from the crew of the *Enterprise*-D. Commander William Riker, born in 2335 in Alaska, could be your great, great, great, great, great, great, great grandson. To put this in perspective, the number of generations from you to William Riker about equals the number of generations from you to William Shakespeare.

If you compare your daily life with that of Shakespeare's time, nine generations seems like forever when it comes to changes in dress, slang, lifestyle, and attitudes. But on evolution's calendar, nine generations is barely the blink of an eye. Natural selection takes thousands of generations to produce major modifications of bodies and brains. That's why our brains and our bodies are very much like those of our seventeenth-century forebears. The physical differences that distinguish us from them come from improved nutrition and health care, not significant genetic changes. If nature is left to its own devices, William Riker and his Starfleet colleagues will have to make do with bodies and brains that are not really any different from ours.

But biological evolution is no longer the paramount agent of human change and adaptation. *Homo sapiens* is a highly social species with a fair degree of intelligence and ability to plan for the future. We exploit our highly developed language skills to share new ideas. Then we alter our own social and physical environments by using technology to put those new ideas into action. And as they pass from mind to mind, our ideas (the "memes" discussed in Chapter 3) can "evolve" in the blink of an eye, spreading like wildfire through populations worldwide. Your whole take on the world can be changed overnight by a book or a television program. Attitudes can be turned topsy-turvy by the siren songs of politicians, scientists, religious leaders, charismatic teachers, or, for that matter, utter crackpots. When it comes to ideas, the human mind is remarkably flexible and, consequently, dangerously gullible.[2]

We humans have been partly liberated from the shackles of biological evolution. Our minds and behavior are shaped by the memes of human invention and human culture. Even in matters of sex, nature is no longer in complete control—Hollywood and Madison Avenue teach us how to win, keep, and shed mates. Our eyes and ears no longer limit our view of reality—inventions such as the electron microscope and the radiotelescope extend our senses into portions of the universe unimagined by our grandparents. Our ability to think and remember is no longer constrained by the neuronal hand that evolution dealt us: artificial brains—computers—make it possible for us to do billions of complex calculations in the time it took Einstein to sharpen his pencil.

If the past is any indication, culturally based changes will continue to pile up at an ever-accelerating pace. Although we can make a guess about the speed of these changes, it's impossible to predict their direction. We're in the same situation as any of our seventeenth-century ancestors who might have speculated about our own time. Never mind the big things, like space flight and particle physics—our ancestors would be just as flabbergasted by some pretty mundane things, including a world in which cities are never dark and in which garbage is recycled. By the same token, we cannot even begin to imagine the worlds our twenty-third-century descendants will inhabit. With all due respect to Gene Roddenberry and his associates, *Star Trek's* portrayal of the future certainly underestimates the changes that will occur.

But this doesn't mean that *Star Trek* is entirely wrong in its predictions. Recent history teaches us that good science fiction—including *Star Trek*—reveals imaginative possibilities, some of which *will* come true. In the nineteenth century, the French writer Jules Verne "invented" vessels for traveling in outer space and under the sea. Verne was uncannily prescient about some twentieth-century scientific achievements. Nearly seventy years ago, in *Brave New World*, Aldous Huxley wrote about the creation of hundreds of genetically identical creatures, including human beings. Huxley called his imagined technique the "Bokanovsky Process." Today we know it as "cloning," but it's no longer just fiction. And in 1945, science fiction writer Arthur C. Clarke speculated about a set of devices that would orbit the Earth and make possible instant global communication. He called these fanciful devices "relays"; their present-day, real counterparts are known as communication satellites.

Think about the technological advances that appear in *Star Trek*.[3] Some of those advances very likely will occur, and when they do, they're going to strain our ethical boundaries and fundamentally alter the way we view ourselves. In this final chapter, we'll focus on advances in three areas: genetic engineering, virtual reality, and telepathy.

Tampering with Nature's Blueprint

Star Trek introduces us to many villains, but Khan Noonien Singh is certainly one of the most depraved of the bunch. During the last decade of the twentieth century, Khan ruled large regions of Earth, including parts of Asia and the Middle East. His brutal dictatorship was eventually overthrown, but only after his tyranny brought death to millions (TOS-1, "Space Seed"; *Star Trek II: The Wrath of Khan*).

Khan is particularly frightening because he is a genetically engineered superhuman. As of this morning's newspaper, medical science hadn't developed all the technology needed to turn out a Khan, or any other superhuman. But stay tuned, because that could change any minute. The blueprints for constructing a human are being worked out even as you read this sentence. In laboratories around the world, an army of molecular biologists is surveying the DNA sequences in each and every one of our genes. This Human Genome Project is labor-intensive and very expensive, but it will soon make it possible for humanity to read its own blueprints.

Of course, once the DNA master plan is worked out, medical scientists will be able to edit and alter our species' blueprints. It will be possible to do a biological cut-and-paste, snipping out an undesirable stretch of DNA and inserting a new, improved sequence in its place. It will be possible to alter the genetic mis-instructions that cause diseases such as cystic fibrosis, or to correct the genetic defects that cause the dozens of forms of hereditary deafness. Want to regenerate part of the eye's retina that has been destroyed by injury? Insert DNA sequences that jump-start the growth process, fooling the adult eye into thinking that it's young again and it's time to grow a retina for the first time. Starfleet's Dr. Toby Russell uses just such a procedure—she calls it "genetronic replication"—to heal Lieutenant Worf's crushed spinal cord following an accident that leaves him paralyzed and suicidal (TNG-5, "Ethics"). Need to correct a malformed heart in a fetus? Extract the erroneous DNA sequence from fetal cells, rewrite the faulty genetic instructions, grow a new heart in the lab, and implant it once the baby is born—or better yet, before it's born. These procedures sound farfetched, but the basic component technologies are already being tested on nonhuman animals.

These forms of genetic engineering are called *genetic therapy*: they cure disease and repair deformity. But the same techniques could be put to work on normal, healthy individuals who want to improve body or mind. Used in this fashion, the techniques are called *genetic enhancement*. Want to forestall the onset of baldness? Simple—reactivate the genetic program that triggers hair growth. Perhaps your memory is already pretty good, but you pine for more CREB protein to make it even better (recall Chapter 6). No problem—insert a new sequence of DNA to regulate the genes responsible for producing CREB.

As *Star Trek* shows us, genetic enhancement can transform a dull-witted boy into a certified medical genius (DS9-5, "Dr. Bashir,

I Presume"). During his early youth, Jules Bashir was frustrated by his inability to keep up with even the slowest of his schoolmates. Stung by guilt over their child's failures, his parents sent Jules off to another planet, where genetic manipulation triggered an explosion of neural growth in his brain. His IQ jumped five points every day for two weeks, and his motor coordination improved dramatically. Out of this manipulation was born a new Bashir—Julian (not Jules any longer, but Julian), a lad destined to become chief medical officer on board Deep Space Nine. When young Julian returned to Earth, he was enrolled in a new school, and his records were falsified to cover up the source of his genius. Bashir became a star pupil, eventually becoming a finalist for the prestigious Carrington Award for outstanding achievement in medical science (DS9-1, "Prophet Motive"). Not bad for someone who began life near the back of the pack.[4]

Of course, when medical technology learns how to improve intelligence by reprogramming DNA, it will have learned how to reprogram other characteristics, too—such as skin color, or sex, or sexual orientation. While these technologies will open new avenues for the treatment of diseases and developmental disorders, they will also challenge the limits of our ethical standards. We're going to have to figure out what constitutes a "disorder." Does mental retardation qualify? If so, what level of retardation justifies genetic repair? What about depression? Short stature? Left-handedness? Reg Barclay-like social phobia? Offensive body odor?[5]

Star Trek reminds us that control of DNA confers an awesome power over others. Just remember the Jem'Hadar: because Dominion genetic engineering left them short one crucial gene, they are enslaved by drug dependency (see Chapter 7). They are unquestioning in their loyalty to the masters who control their drug supply (DS9-3, "The Abandoned"; DS9-4, "Hippocratic Oath"). The Jem'Hadar are genetically designed drug addicts.

One form of genetic engineering making headlines these days is cloning. As explained in Chapter 3, cloning uses a single individual's DNA to create a new individual. Actually, cloning is nothing new. Every time you cultivate a cutting from your favorite begonia, you perform an act of cloning. Whether the clone is plant or animal, however, it looks identical to its DNA donor because both are made from the same set of genetic blueprints.

Cloning plants is one thing, but cloning human beings is quite another. The existence of cloned humans will create a host of difficult ethical issues. To head off these thorny issues while they're still in the theoretical stage, governments have already begun to pass laws against cloning humans. The Council of Europe, for example, recently prohibited "any intervention seeking to create a human being genetically identical to another human being, whether living or dead."[6] This peculiar precautionary prohibition was instigated by one dramatic step toward cloning our own kind: the first successful cloning of a mammal.[7]

Ian Wilmut and a team of other Scottish mammal makers began by removing a cell from an adult sheep's mammary gland. Next, they took an egg from a second sheep, sucked out the egg's DNA, and replaced it with the DNA from the first sheep's mammary cell. This hybrid egg began dividing, as normal eggs do, and soon formed an embryo. The embryo was then implanted into a surrogate mother, who eventually gave birth to a lamb affectionately named "Dolly."

Today, Dolly is a perfectly normal, unremarkable animal in all ways but one: her DNA exactly matches that of the adult sheep whose cellular donation started the ball rolling. Does identical DNA mean that Dolly is in every way a perfect copy of her DNA donor? No. As biologist Richard Lewontin reminds us, "It takes more than DNA to make a living organism."[8] Dolly is not reducible to her DNA—and neither are you.

Just look in the mirror. The left and right sides of your face are not exactly identical, even though the same genes contributed to the construction of both sides. Some of these subtle side-to-side differences came from random molecular movements at the time your fetal face was being formed. Such random movements—developmental "noise"—introduce inevitable variations in the way a DNA sequence is expressed.

The environment in which DNA operates also influences its expression.[9] A good example is the cozy relationship between genes and hormones. At a crucial moment during the development of a male fetus, a gene stimulates the production of the hormone testosterone. An enzyme converts testosterone into a closely related substance, dihydrotestosterone, which turns on other genes that promote the growth of a penis and scrotum. At this moment, the fetus is on its way to becoming a bouncing baby boy. In rare cases, though, something derails the process. A genetic male (i.e., a fetus whose cells have the normal male chromosome pattern) fails to produce the enzyme that converts testosterone to dihydrotestosterone. And no dihydrotestosterone means no penis, no scrotum. The genes that promote the development of the external genitals are not turned on, and these body parts fail to develop. With no evidence to the contrary, the parents may raise the child as a girl. But puberty brings some surprises. The *internal* male sexual apparatus, which this individual does have, suddenly spews out testosterone. This testosterone surge triggers the development of a penis, deepens the voice, and produces a characteristic male physique.[10]

The hype of the Human Genome Project sometimes eclipses a simple, important principle: Genes depend on environments for their expression. And those environments are shaped by the activities of the organisms that live in them. To quote Lewontin again, "Organisms do not experience environments. They create them."[11] It seems obvious that the environment in which Dolly the cloned lamb grows up will shape her development. What may be less obvious is that Dolly's own behaviors will be an important part of that environment. By her eating habits, her interactions with other animals, and her levels of stress, Dolly herself will create the environment she experiences during her journey through life. Dolly's genes are just one part of the luggage she carries on that journey.

Ditto for you. The actions of your genes depend on the experiences you've enjoyed (and suffered), particularly during the formative stages of development in the first two decades of life. The world in which you grew up was defined by your parents, peers, teachers, and even strangers. Whether it's stimulating or stultifying, the local world that adults create for a child greatly influences that child's development. That world determines how much of the child's genetic potential will come to fruition—and how much will be wasted.[12]

The genetic engineering story has taken an interesting twist. By building a brain capable of consciousness and invention, our genes have allowed us to create and use tools to modify the very entities—the genes—responsible for the design of those tools. Armed with those tools, we can now sidestep nature's patent on the methods for creating the organisms that populate the earth's environment. Our genes, in other words, have lost some of their power to determine our fates. *Humans* are in control here.

In Thine Own Image

Cloning can create real, genuine, flesh-and-blood creatures—and someday, real people with real personalities. *Star Trek* takes the engineering challenge one step further: three-dimensional projection techniques produce high-fidelity virtual environments and virtual people practically indistinguishable from the real thing. Thanks to holographic technology, Starfleet personnel can create virtual people with virtual personalities inhabiting virtual environments.

Galaxy-class starships like the *Enterprise*-D routinely include a holodeck—technically, a holographic environment simulator. Crew members use the simulations on these first-generation holodecks for exercise, relaxation, and job training.[13] On board Deep Space Nine, the Ferengi barkeep Quark exploits holographic technology for commercial purposes. Judging from the station's scuttlebutt, sexual fantasy programs are the hot attraction in Quark's holosuites (DS9-3, "Meridian"). But Quark's holosuites offer more than just sex and more sex; other popular virtual activities include kayaking (DS9-3, "Heart of Stone"), reenactments of historic Klingon battles (DS9-2, "Blood Oath"), and instant replays of historic sporting events, including that defunct game called baseball (DS9-1, "If Wishes Were Horses"; DS9-4, "Starship Down").

Holographic technology reaches its zenith aboard the U.S.S. *Voyager*: a key member of the crew, the ship's doctor, is a holographic projection. *Voyager's* crew members establish close relationships with this virtual physician and treat him as they would any other humanoid with a medical degree.[14] The wry humor of this virtual human is as excellent as his medical skills.

It may be a while before you have your own resident virtual physician. Still, the seeds of sophisticated holodeck technology are already sprouting in laboratories around the world. By *Star Trek* standards, the simulations produced by existing devices are Stone Age, but virtual environments are being used effectively in education, entertainment, and medicine. They are even being used in psychotherapy to treat phobias, one of the mental disorders discussed in Chapter 7.

The treatment of phobias begins with the recognition that phobias are learned, not innate. If phobias can be learned, then they can be unlearned. But where do we begin the unlearning? Suppose that someone suffers from acrophobia (fear of heights). It does no good to whisk him to the top of the Empire State Building, expecting to convince him that his fears, like his feet, are ungrounded. The treatment must proceed in small steps. With someone who is afraid of heights, the therapist works out a list of situations that produce varying degrees of anxiety in the patient— from standing on the first rung of a sturdy ladder, all the way up to standing on the ledge of a tall building. The patient tries to visualize himself in each of these situations, starting with the least fear-provoking one and working his way up.

If this technique is to succeed, the visualization must be vivid, and not everyone is good at this. A therapeutically valuable alternative is to put the patient into a highly realistic simulation. Now, no visualization is needed; the patient merely has to keep eyes and ears open to virtual reality. Such virtual environments make it possible to present realistic, carefully controlled settings for therapeutic desensitization. Acrophobic patients, for example, can view anxiety-producing scenes from virtual bridges, virtual balconies, and a virtual glass elevator. Though their feet never really leave the ground, the experience is compelling. Beginning at ground level, the patients gradually move higher in each simulated scene. Upon reaching each new level, the patients remain at that "height" until their anxiety subsides.[15] Then, with new confidence and calm, they push upward to the next level. Within a few weeks, the patients' high anxiety falls dramatically. Comparable systems that treat fear of flying are being tested now—soon, the white-knuckled flier may be a thing of the past. Virtual environments have also been used successfully to reduce arachnophobia (fear of spiders)[16] and several other phobias.

Despite its successes, present-day virtual reality has barely scratched its virtual surface. Just imagine what can be done once we can create realistic simulations of human social interactions. We'll be able to experience simulated give-and-take with highly realistic holographic representations of all sorts of people who are significant in our lives. A real therapist will be able to enter the simulation and guide the patient through the experience. A young child will be able to confront—in a safe, controlled setting—that bully who's been tormenting him in the schoolyard. These virtual confrontations will teach the child how to deal with his tormentor (and other tormentors, later in life). Or suppose you want to repair an important personal relationship that's turning sour. Using virtual reality technology, you can interact with a realistic simulation of your friend or lover, trying out different styles of interaction. You'll even be able to interact with significant others who are no longer living, working out psychological hangovers from those old relationships. If current successes are an indication, this technology will help ease social phobias and sexual hangups.[17]

Virtual reality will let us see ourselves as others see us. Not too long from now, therapist-programmers will create simulations in which clients watch their own social interactions from a third-party perspective. You'll watch your own likeness deal with other people, and you'll understand why people behave toward you as they do. Each partner in a troubled relationship will be able to experience the world (and that relationship) from inside the virtual body of the other partner—the ultimate in walking in another person's shoes!

These possibilities are intriguing, but they carry a hefty price tag. When virtual reality technology advances to *Star Trek's* level, we'll be faced with an enormously complex ethical issue: How do we treat people who seem real but aren't? When virtual people are temporary bit players, they exert a limited pull on the real people they come in contact with. If virtual acquaintances are killed or injured, there's not much concern about their erasure from existence. But virtual friends are a different matter. It's possible to get emotionally too close to a holographic simulation. That's exactly what Commander Riker discovers when he falls head over heels for Minuet, a woman he meets in a holodeck simulation of a New Orleans jazz bar (TNG-1, "11001001"). Unfortunately for Riker, the virtual vamp turns out to be the holographic creation of advanced humanoids temporarily visiting the *Enterprise*-D—upon leaving, they erase her program and, thus, dash Riker's dreams. The same fate befalls *Voyager's* Tuvok, who becomes infatuated with a holodeck female named Marayna (DS9-3, "Alter Ego"). Drawn by her "impeccable logic," Tuvok yearns to spend lots of time with her, a novel urge for this usually well-buttoned-up Vulcan. But, again, the ending isn't happy. Marayna turns out to be the creation of a hideous alien trapped in a lonely existence. Tuvok quickly loses interest, impeccable logic or no impeccable logic.

When we spend lots of time with a virtual person, that person will take on a reality of its own, both in its "virtual" mind and in the real minds of others. This is precisely the situation Odo and Jadzia Dax discover when they visit a quaint village on a planet in the Gamma Quadrant (DS9-2, "Shadowplay"). There they meet Rurigan, a man who escaped when his homeworld was conquered by the Dominion. Rurigan has settled on this new planet and has used a hologenerator to recreate his home village, complete with buildings and people. This replica is very realistic: the virtual people laugh, cry, get married, and have children, and they exhibit what can only be described as true sentience. But the hologenerator develops instabilities and must be shut down for repairs. After the shutdown, only Dax, Odo, and Rurigan remain. When Rurigan decides to terminate the holographic world he has created and return to his home planet, Odo is stunned. What about the future of the to-be-terminated villagers, including that cute little girl who is Rurigan's virtual granddaughter?

RURIGAN: She's still a hologram.

ODO: Maybe. But I saw the way you held her hand when she was sad. I saw the way you tried to comfort her when she was frightened.

RURIGAN: I didn't want her to get hurt.

ODO: If she's not real, what does it matter?

RURIGAN: It matters—it matters to me.

ODO: Why should it matter to you if a hologram cries?

RURIGAN: Because I love her.

DAX: And she loves you.

ODO: Don't you see? She's real to you. And she's real to me, too. They're all real. And you can't turn your back on them now.

Any twentieth-century parent who has watched a child become attached to a doll or a stuffed animal knows that Odo is right. Once you've established a rewarding relationship with another—whether it's another person, an animal, or even an object—you're hooked. This is something we'd better keep in mind as holographic technology brings us to the threshold of creating virtual people. Virtual friends and relatives will become genuine realities in our lives. They'll be easy to turn on, but heartbreaking to turn off.

A Penny for Your Thoughts

Betazoids are known throughout the universe for their telepathic abilities. Avoiding all local and long-distance charges, they send and receive thoughts mentally. Virtual reality and genetic engineering may be just over the horizon, but telepathy is one ability that may never be mastered by our species. Or perhaps it already has been, depending on how you look at it.

Should we humans be envious of Betazoids? After all, telepathy does seem like a pretty handy skill. You could use it to tell whether someone was genuinely happy or just pretending to be, and it could tell you—without your having to ask—what was on a friend's mind. There are many ways that telepathy could make your daily life easier and safer and more entertaining.

Deanna Troi's telepathic ability makes her a better counselor to the crew of the *Enterprise*-D. At one time or another, each of the ship's officers benefits from the insights extracted by Troi's penetrating mind. In addition, Troi routinely serves as Captain Picard's humanoid "barometer," reading the thoughts and feelings of others to ascertain who is lying and who is telling the truth. A slight, telltale nod of Troi's head is all the signal that Picard needs.

Telepathy expands Troi's sense of reality, as we learn when she temporarily loses her empathic abilities (TNG-4, "The Loss"). An array of two-dimensional life-forms accidentally crosses the *Enterprise*-D's trajectory and begins dragging the ship along in its wake. The presence of these anomalous beings robs Troi of her telepathic powers. She can't feel the psychic presence of her crewmates. Everyone around her seems hollow and empty, drained of emotional color. Devastated by her "psychic blindness," Troi resigns her position as ship's counselor. Fortunately for her, and for the *Enterprise*-D, the life-forms abandon their hold on the ship, thereby restoring Troi's expanded psychic reality.

Telepathic ability, however, is not all fun and games and being ship's counselor. It also creates problems for its possessors. For one thing, the channel it opens between minds tends to be two-way. According to Mr. Spock, nearly all of the universe's known telepathic species automatically broadcast their own thoughts for others to read (TOS-3, "The Empath"). A telepath's mind is as open as a book can get. Poor Deanna Troi is reminded of this every time her overbearing, flamboyant mother comes to visit. Lwaxana Troi has the annoying habit not only of eavesdropping on Deanna's thoughts, but also of giving her unwanted motherly advice—telepathically, of course.

Captain Jean Luc Picard and Dr. Beverly Crusher, ordinary humans with no telepathic abilities, temporarily experience telepathy's two-way traffic. Taken prisoner on an alien planet, they receive neural implants that can extract strategic information from their brains (TNG-7, "Attached"). After Picard and Crusher escape, they discover that each one's thoughts are being "heard" by the other. Their mental duet gets confusing.

CRUSHER: So am I.

PICARD: What?

CRUSHER: I'm thirsty too.

PICARD: I didn't say anything about being thirsty.

CRUSHER: I heard you. You said, "I'm very thirsty."

Each person's neural implant picks up and transmits its owner's thoughts, injecting them—uninvited—into the mind of the other. Although it's not done maliciously, this invasion of privacy is disturbing. Later in the day, Crusher is hungry, and naturally her thoughts turn to food. Picking up the message, Picard complains that *one* of them must be hungry. Crusher confesses that the thought of a bowl of vegetable soup belongs to her. Picard begs her to think of something else—*her* thoughts are making *him* hungry. Telepathic contagion.

During their temporary mind-meld, Picard and Crusher discover another nuisance that accompanies the "gift" of telepathy. Intrusive "mind noise" from other people constantly disrupts the telepath's own thoughts.

> **PICARD:** Isn't it astonishing, though, how much clutter there is in the consciousness? Odd memories coming to the surface, bits of half remembered songs, . . .
>
> **CRUSHER:** [continuing Picard's sentence] . . . stray daydreams, scattered minutiae. Wonder how true telepaths sift through it all? How can they really get to what someone's thinking if the mind keeps churning all this flotsam to the surface?

Crusher has put her finger on something very important. It's hard enough to sort through the torrent of sensory information that comes from our eyes, ears, and other senses. Add to this a stream of telepathic input and your mind would be swamped. Just think back to the last time you were waiting at a crowded airport for the arrival of a friend. The bustle of people and the cacophony of voices made it difficult to spot your friend or to hear her calling your name from across the waiting area. But suppose at the same time, you were bombarded by an avalanche of thoughts and feelings from every person in the terminal. This telepathic input would exceed your attentional capacity. Individuals who have powerful telepathic talents would live in a continuous mental din, bombarded with telepathic "noise" from the minds of others. It's no surprise that Betazoids with particularly acute telepathy can get so stressed out that they seek respite in a psychiatric hospital (TNG-3, "Tin Man").

And what about the real possibility that it might be wrong to use telepathy? Are there particular times when telepathy would be unethical? Deanna Troi is very careful to restrict her telepathic activities to business. When she interacts socially with nontelepaths, she relies on their spoken words and visible expressions, and she never,

ever, cheats at poker by reading the minds of others around the table. But not all telepaths are so scrupulous.

Devinoni Ral, a skilled negotiator, is one-quarter Betazoid, enough to give him telepathic abilities (TNG-3, "The Price"). But others don't realize that he's a telepath, making it possible for him to exploit his powers for material gain. Deanna Troi accuses Ral of dishonesty because he uses telepathically acquired information to manipulate people. Ral counters by pointing out that people have been reading each other's minds for thousands of years: they watch each other's actions and they listen carefully.

Devinoni Ral is right. Members of our human species are indeed pretty good at what could be characterized as "mind-reading." But—and this is a *very* important "but"—we don't literally read the thoughts from another person's mind, nor do we directly sense someone else's feelings.[18] We *do* possess an extraordinarily keen blend of perception, intuition, and empathy. Together, these natural talents endow us with mental powers that rival telepathy.

How do we do it? By picking up myriad, subtle cues to other people's thoughts and feelings: facial expressions, voice intonation, body language. And we do it without even thinking. Your wife insists that she's not disappointed about the canceled vacation, but her face tells the truth. When you talk on the phone to your son at college, the excitement in his voice automatically tells you that he's aced his first exam. When your eyes light up at the beautiful red convertible in the auto dealer's showroom, the car salesman knows he's got a hot prospect—no matter how nonchalant you try to seem.

In countless social situations, humans behave like mind-readers, using cues to formulate a theory about what the other person is thinking and feeling. And going one step further, we use that theory to predict how that person will behave. We formulate a "theory of mind"—an explanation of what makes somebody else tick. We create theories of mind whenever we want to anticipate the actions or account for the motives of another person.[19]

To illustrate how theory of mind works, Angeline Lillard invited readers to consider the tale of Little Red Riding Hood.[20] Here's her bare-bones, stripped-down version of the classic story:

A little girl hears from a woodcutter that her grandmother
is sick. She walks to her grandmother's house, carrying a
basket of treats. A wolf who is in her grandmother's bed
jumps out and runs after the girl.

Notice that this account says nothing about the characters' motives. Why did the woodcutter tell Little Red Riding Hood that her grandmother was sick? What was on her mind when Little Red Riding Hood walked to granny's house? And how about that wolf? What on Earth was he doing in granny's bed, and why did he chase Little Red Riding Hood? People can answer most of these questions even if they've never heard the whole story before. Their "theory of mind" allows them to make inferences about the characters' motives and feelings.

We tend to assume that the crimson-clad heroine set out for her granny's house because, as a loving grandchild, she was concerned about her granny's health and wanted to comfort her. A plausible explanation—but not the only one. If a Ferengi child heard the stripped-down version of the story, that child would automatically assume that the little girl's trip was motivated by a desire to claim and sell her granny's remains. After all, that is the custom when Ferengi die (DS9-2, "The Alternate"; DS9-4, "Body Parts"). In contrast, a Klingon child would naturally assume that Little Red Riding Hood went to her grandmother's house out of a sense of family honor and pride. And that Klingon child would expect Little Red Riding Hood to defy the wolf's challenge, stand her ground, and fight for her honor, knowing that if she lost it would be a good day to die.

Sometimes our theories of mind are just plain wrong. With absolute sincerity, a wife asks her husband, "Do you think the oil needs changing in the car?" He mistakes her comment to mean that the oil *does* need changing *and* that he should have taken care of it long ago. He storms out of the house, leaving her to figure out why her innocent question angered him. His theory of her mind was wrong. And it's not just husbands and wives who make mistakes as mind readers.[21] A mother reminds her teenage son to turn off the hall light when he gets home; he accuses her of not trusting that he'll be in by midnight. An employee asks if he can do anything else before heading home; his supervisor mistakenly takes this to mean that the employee is eager to leave.

Some individuals constantly misread the minds of others. We call one group of such individuals "paranoids." These highly suspicious people routinely question the motives of others and read hidden meanings into their innocent remarks and actions. Their theories of mind are pathologically wrong. At the other extreme, some individuals are real pros at forming accurate theories of mind—they're remarkably adept at sensing what others might want or might be feeling. Unfortunately, some superlative mind-readers exploit their mind-reading skills by manipulating people for their own personal gain. Such individuals are labeled con artists or, in the extreme, sociopaths.

Finally, some individuals are unaware that other people even have thoughts and feelings of their own. These social agnostics simply fail to form theories of mind, accurate or inaccurate. *Voyager's* Vulcan security officer Tuvok sometimes appears to have this problem. Despite his keen, dispassionate intellect, Tuvok lacks empathy and interpersonal skills.[22] Take Tuvok's behavior when he and his *Voyager* crewmate Neelix are trapped in a carriage on an orbital tether (VOY-3, "Rise"). The carriage's uncontrolled acceleration is about to kill its occupants. Neelix comes up with a way to avert the impending disaster, but Tuvok curtly dismisses Neelix's idea. This infuriates Neelix, who accuses Tuvok of having no instincts, no empathy, and no understanding of people.

Neelix's plan does, in fact, save the carriage and its passengers. Does Tuvok reward Neelix with praise and admiration? Afraid not. True to his Vulcan heritage, Tuvok confirms his lack of empathy as he puts Neelix down, hard:

> *Your instincts were correct. However, one day your intuition will fail and you will finally understand that logic is primary above all else. Instinct is simply another term for serendipity.*

Vulcans aren't the only individuals who have difficulty forming theories of mind. Among twentieth-century humans, there are people who seem to find it impossible to view the world from another person's perspective. These individuals suffer from autism, a complex, cruel disorder characterized by impaired interpersonal skills. Autistic individuals avoid eye contact with others, prefer to be alone rather than with others, and have extreme difficulty judging other people's moods or intentions.

Among autism's victims is Dr. Temple Grandin, the writer and world-renowned veterinarian who was introduced in Chapter 1. Her Vulcanlike symptoms underscore the real paradox of this disorder: autistic individuals can be intelligent and highly creative, yet socially and empathetically backward. They have trouble understanding what makes other people tick—they can't formulate the "theories of mind" that usually make our social interactions easy, natural, and predictable.[23] Temple Grandin has said that Spock's behavior and his separateness from the other crew members helped her understand her own condition. We bet that Gene Roddenberry would have been very pleased.

Will humans always have to rely on theories of mind to tell us what others are thinking and feeling? Or will we someday be like Betazoids, able to send and receive thoughts telepathically? Human thought consists of patterns of neural activity distributed throughout the brain. When you are thirsty and think about a cold beer, clusters of neurons are active in several parts of your brain, including the limbic system and frontal cortex. Suppose we could record those patterns of electrical activity as they were occurring, route the signals to a tiny broadcasting device, and transmit them to a receiver implanted in a friend's brain. If that receiver were able to distribute the signals into the appropriate regions of your friend's brain, those signals could trigger images of a cold beer in the mind of that individual.

Sound farfetched? Many *very* high technical hurdles would have to be overcome. But those hurdles may not be insurmountable. Work already under way shows that it's possible to pick up electrical activity from an animal's brain and use those signals to operate a robotic arm. In one such study, the neurons being monitored are located in the portion of a rat's brain that controls the animal's movements, the so-called motor cortex. The animal is trained to produce patterns of activity in its motor cortex that move a robotic arm toward its mouth, so that it can obtain a food reward carried by the arm.[24] (The real aim of this work is to develop implants that will allow people paralyzed by spinal injury to use their functioning brains to control the movements of surrogate arms and legs.)

But can we route electrical signals into the brain of another person? And can those second-hand signals produce meaningful consequences? Here again, the answer is yes. Neurologists have already constructed devices that artificially activate brain cells. As mentioned in Chapter 5, one such device has been implanted in the brains of blind people. Tiny electrical currents passed into the visual cortex trigger visual sensations—the person wearing the device "sees" flashes of light in his visual field.

Make no mistake about it: we're still a long way from being able to monitor activity in the hundreds of millions of brain cells that participate in the generation of complex human thought. And once that monumental problem is solved, transmitting those signals to another person's brain and distributing them to the correct areas will present even greater challenges. So it's unlikely that anything close to a neural telepathy implant will be available during our lifetimes. But don't forget: Our seventeenth-century ancestors would have laughed at the idea that you could broadcast visual images through the air, making it possible for people all around the world to watch Klingons, Cardassians, or even Cosmo Kramer. So before you dismiss the idea of a neural telepathy device, keep in mind Lieutenant Commander Data's wise observation: "All things which can occur, do occur" (TNG-7, "Parallels").

A Final Thought

"Five card stud, nothing wild, and the sky's the limit." With these simple words, Jean Luc Picard closes the final episode of the seventh and last season of *Star Trek: The Next Generation* (TNG-7, "All Good Things . . . "). During that episode, Picard once again earns his Starfleet pay, stopping an unusual backward-spreading temporal anomaly and saving humanity. Q congratulates the captain of the *Enterprise*-D and explains why the Q-continuum created this anomalous anomaly:

> We wanted to see if you had the ability to expand your mind and your horizons . . . That is the exploration that awaits you. Not mapping stars and studying nebulae, but charting the unknown possibilities of existence.

Q was talking to Picard, but he was also talking to us. And we should listen.

Who's Who and What's What

BARCLAY, REGINALD (TNG). Human. Socially challenged engineer on *Enterprise*-D. Nicknamed "Broccoli" by his crewmates. Has rich social and love lives, but only in holodeck simulations. Has a morbid (and sometimes justified) fear of using a transporter.

BASHIR, JULIAN (DS9). Human. Medical officer of Deep Space Nine. Genetically enhanced in childhood to make up for his subpar intellect. Close friend and darts-playing companion of Miles O'Brien.

BORG (TNG, VOY). Humanoid-machine hybrids. Individual Borg's nervous systems are linked together to share knowledge and produce a group consciousness. The Borg collective expands its territory and technology by assimilating other, more technologically advanced species. Fond of warning potential victims, "You will be assimilated; resistance is futile." This warning is usually, but not always, correct.

CARDASSIANS (TNG, DS9). Humanoid species, historically allied with the Dominion. Occupied Bajor during the twenty-fourth century. Builders of the space station known as Deep Space Nine (Cardassian name: Terok Nor).

CHAKOTAY (VOY). Human. First officer aboard *Voyager*. Proud of Native American heritage. Distinguishing marks include, on left temple and forehead, a tattoo worn to honor his father and ancestors.

CRUSHER, BEVERLY (TNG). Human. Medical officer aboard *Enterprise*-D (except when temporarily replaced during Season 2 of TNG). Widow of Starfleet officer Jack Crusher and mother of Wesley Crusher. Close confidant of Captain Jean-Luc Picard.

CRUSHER, WESLEY (TNG). Human. Teenage son of Beverly Crusher and the late Jack Crusher. Technical whiz; precocious. Eventually leaves Starfleet to explore space/time dimension.

DATA (TNG). Android. Operations officer on *Enterprise*-D. Designed by Noonien Soong in his own image. Externally, an anatomically correct human male, except for distinctive alabaster skin and yellow eyes; internally, made up of electronic chips and a capacious positronic brain. Has violent sibling rivalry with older, evil brother Lore. Entirely rational, except for an interlude when an emotion chip is installed. Aspires to understand human beings, a goal most humans also aspire to.

DAX, JADZIA (DS9). Trill. Female humanoid host of a vermiform symbiont. This symbiont, named Dax, has had seven prior hosts of both sexes. Jadzia has access to memories accumulated during her symbiont's previous incarnations. Immediate predecessor host, Curzon, was a close male friend of Benjamin Sisko, who is now Jadzia Dax's commanding officer. Referring to the symbiont's prior host, Sisko affectionately calls Dax "Old Man."

DEEP SPACE NINE (DS9). The name Starfleet uses for the mining station originally built by the Cardassians near the planet Bajor in the Alpha Quadrant. Deep Space Nine occupies a strategic position near the opening of a stable wormhole (a tunnel through subspace), which permits rapid movement between the Alpha and Gamma Quadrants. Captain Benjamin Sisko oversees the space station under the joint auspices of the Bajoran government and Starfleet.

DOMINION (DS9). An empire based in the Gamma Quadrant. Established by shape-shifters in response to persecution and widespread anti-shape-shifter prejudice. As their empire grew, the shape-shifters became known as the Founders of the Dominion ("The Founders" to their friends). The Dominion's administrators, the Vorta, worship the Founders as gods.

EMERGENCY MEDICAL HOLOGRAM (VOY). Holographic projection. Has no genuine name (aside from "Kenneth," as he is called by his short-lived holowife). Created in the image of Dr. Lewis Zimmerman, Starfleet holoengineer. Initially designed for temporary, backup medical duty; his memory contains all of Starfleet's medical knowledge. Fond of opera; accomplished amateur tenor. Wry sense of humor. Social skills and bedside manner need some fine-tuning: "You have a lovely brain. It will make a fine addition to our files."

ENTERPRISE (TOS, TNG). Name given to one United States nuclear-powered aircraft carrier in the twentieth century, and to six different United Federation of Planets starships, beginning in the twenty-third century. The best known of these are the U.S.S. *Enterprise*, a *Constitution*-class vessel commanded by several captains, most notably Captain James Tiberius Kirk, and the U.S.S. *Enterprise*-D, a much larger, *Galaxy*-class vessel commanded by Captain Jean-Luc Picard.

FERENGI (TNG, DS9). Humanoid species of short stature, with very large ears, jagged teeth, and large, bulbous noses. Homeworld is

Ferenginar, a planet with nearly constant rain. Ferengi society's wheels are powered by avarice gone wild. Folk wisdom is embodied in the Ferengi Rules of Acquisition. The crash on Earth of a Ferengi spacecraft in August 1947 triggers persistent reports that aliens landed near Roswell, New Mexico.

GARAK, ELIM (DS9). Cardassian. Onetime high official in the Obsidian Order (the Cardassian security agency), now runs a tailor shop aboard Deep Space Nine. Suspected of being a Cardassian spy. Takes pride in ability to devise creative lies: "The truth is usually just an excuse for lack of imagination."

GUINAN (TNG). El-Aurian. Barkeep in *Enterprise*-D's bar/lounge, Ten-Forward. Her species was almost entirely exterminated by the Borg. Source of sage advice for crew of *Enterprise*-D, especially Captain Picard.

HOLODOCTOR (VOY). *See* **EMERGENCY MEDICAL HOLOGRAM.**

JANEWAY, KATHRYN (VOY). Human. Captain of the U.S.S. *Voyager*, a starship stranded in the Delta Quadrant of the Milky Way galaxy when it is blown way off course. Pursues artistic projects in holodeck re-creation of Leonardo da Vinci's workshop. Undaunted by the distance and the dangers of the Delta Quadrant, Janeway is determined to get her crew home, no matter who or what stands in her way: "I don't like threats, I don't like bullies, and I don't like you!"

JEM'HADAR (DS9). Genetically engineered humanoid species. Loyal, obedient, and brutal. Enforcers for the Dominion. Engineered with an enzyme missing; require constant infusion with the drug ketracel to stave off anxiety. Depend on the Vorta, administrators for the Dominion, for ketracel. Reach maturity very rapidly; grown up and out of the house almost before you know it.

KES (VOY). Ocampa. Nurse and friend to *Voyager's* Emergency Medical Hologram. Ocampa life expectancy is equivalent to seven Earth years. Ocampa mature rapidly, and females have just one, brief opportunity to reproduce. Has some telepathic ability. Eventually leaves *Voyager* to explore disembodied spirituality.

KIM, HARRY (VOY). Human. Ensign aboard *Voyager*. Extremely unlucky: nearly eviscerated by DNA-hungry vixens; locked in a brutal prison where he was beaten and starved; tormented by a malicious, computer-generated clown; panicked when Seven of Nine ordered him to take off his clothes for sex.

KIRA NERYS (DS9). Bajoran. Second-in-command of Deep Space Nine. Formerly a member of the Bajoran militia and combatant in the guerrilla war of liberation against the Cardassian occupation of Bajor. Surrogate mother of Miles and Keiko O'Brien's child Yoshi.

KIRK, JAMES T. (TOS). Human. Born on Earth in Iowa. Captain of the original *Enterprise* and its first successor, *Enterprise*-A. Attains rank of admiral. Survivor of massacre on Tarsus IV, corporeal bifurcation during transporter malfunction, thousand-foot fall while mountain climbing, court martial, murder conviction, and involuntary exchange of bodies with ex-girlfriend. Easily distracted by an attractive female. Described by one enemy as a "swaggering, overbearing, tin-plated dictator."

KLINGONS (TOS, TNG, DS9). Combative members of a society governed by warrior code and arcane rituals. Klingon opera is a favorite musical entertainment, though hard on non-Klingon ears. Language, Klingonese, is extremely guttural; spoken on Earth today by about one thousand human admirers of the language (for instruction, Visit the Klingon Language Web site at **http://www.kli.org/**). Favorite intoxicating beverage: blood wine.

LAFORGE, GEORDI (TNG). Human. Chief engineer of *Enterprise*-D. Mother was a Starfleet officer. Born blind. Equipped at age six with sophisticated visual prosthesis, VISOR (Visual Instrument and Sensory Organ Replacement), which allows expanded "sight" over a wide portion of the electromagnetic spectrum. Close friend of the android Data.

McCOY, LEONARD H. (TOS). Human. Medical officer of the original *Enterprise*. "Choleric" is the word he brings to mind. Nicknamed "Bones" by Captain Kirk. Fond of justifying his inability to fix some problem by reminding listeners that his practice is limited: "I am a doctor, not a bricklayer." "I am a doctor, not an escalator." "I am a doctor, not a coal miner."

NEELIX (VOY). Talaxian. *Voyager's* self-appointed cook and morale officer: "Your voice says to go away, but your heart wants me to make you smile!" Ship's resident expert on the Delta Quadrant of the Milky Way galaxy. Colorful, unruly hair; by human standards, every day is a bad hair day for Neelix.

NOG (DS9). Ferengi. Son of Rom. Nephew of Quark. Friend of Jake Sisko. Hyper enthusiastic Starfleet cadet. Chronically suspicious of the Cardassian Garak.

O'BRIEN, MILES (TNG, DS9). Human. Chief engineer aboard Deep Space Nine. Close friend and darts-playing pal of Julian Bashir. Husband of Keiko and father of Molly and Yoshi. Fiercely proud of Irish heritage and culture.

ODO (DS9). Founder. Security chief on Deep Space Nine. Like other Founders, a liquid life-form able to take on appearance of any solid object or life-form. Sometimes described as a "shape-shifter" or "changeling." Dies if he fails to return to original, liquid form once every sixteen hours. Usually takes a form that approximates a humanoid male. Eternally suspicious and scowling; especially watchful for missteps by Quark or Garak. Sleeps in a bucket.

PARIS, TOM (VOY). Human. Onetime Starfleet officer, defected to Maquis. Captain Janeway arranged his release from a penal colony so that he could take on temporary duty aboard *Voyager*. *Voyager's* mishap means that Paris's duty will be anything but temporary. Friend of Harry Kim. Romantically involved with B'Elanna Torres.

PICARD, JEAN-LUC (TNG). Human. Captain, *Enterprise*-D. As a result of a bar fight, has an artificial heart. Once captured by the Borg and partially assimilated; in his assimilated form, Locutus of Borg, helped the Borg defeat a Starfleet armada at Wolf 359. Fond of quoting Shakespeare, but not in the original, Klingon version. Favorite beverage: Earl Grey tea, hot. Romantically, far more discreet and self-controlled than Kirk.

Q (TNG, DS9, VOY). Designates both (1) a collection of nearly omnipotent and quasi-omniscient beings, the Q-continuum, and (2) any individual member of that collection. Of the individuals, one is fond of elaborate but cruel practical jokes at the expense of humans, presiding over kangaroo courts, and cross-dressing in Starfleet uniforms. Q relishes the chance to introduce the crew of the *Enterprise*-D to the Borg.

QUARK (DS9). Ferengi. Proprietor of Quark's Place, Deep Space Nine's bar and gaming establishment; also proprietor of the station's holosuites. At once both critical and protective of his less bright brother, Rom. Least favorite beverage: root beer, which is too bubbly and cloyingly sweet for his taste, much like the United Federation of Planets. Flirtatious behavior borders on sexual harassment.

RIKER, WILLIAM (TNG). Human. First officer, *Enterprise*-D. Son of a former Starfleet officer who is now a civilian advisor to Starfleet. Formerly Deanna Troi's college boyfriend. Enjoys reputation as a ladies' man. Passable trombone player. Has a twin brother Thomas Riker, created during a transporter accident.

ROM (DS9). Ferengi. Slow ("couldn't fix a bent straw") but well-meaning brother of Quark. Onetime engineer in Starfleet. Very fond of his mother, whom he affectionately calls "Moogie." Married to Leeta, a Bajoran female who works at Quark's bar.

ROMULANS (TOS, TNG, DS9). Homeworlds: Romulus and Remus. Passionate and aggressive race that split off from Vulcans several thousand years ago. Their alliance with Klingons disintegrated into bitter war, leaving the two mortal enemies. Famous for their highly intoxicating light blue ale, which is illegal in the Federation.

SAREK (TOS). Vulcan. Vulcan ambassador and estranged father of Spock. Disappointed that his son didn't go into his business (diplomacy), but chose to work for Starfleet instead.

SCOTT, MONTGOMERY (TOS). Human. Chief engineer of the original *Enterprise*. Called "Scotty" by nearly everyone. Often exasperated by the gap between his ship's capabilities and his captain's insistence that they go faster; longs for ability to repeal the laws of physics.

SEVEN OF NINE (VOY). Human female captured as a child by the Borg and assimilated into their collective. While still Borg, she was captured by the *Voyager* crew and deprogrammed. Has been assimilated into the *Voyager* family and serves as assistant to Chief Engineer Torres. Displays remnants of Borg behavior and wisdom: "Oral consumption is inefficient." Ensign Kim (along with quite a few others) finds Seven attractive. She's known to her friends on *Voyager* as plain old Seven.

SISKO, BENJAMIN (DS9). Human. Captain, Deep Space Nine. Before his posting to Deep Space Nine, wife Jennifer was killed, during Borg attack on Wolf 359. Baseball enthusiast, even though the game is extinct. Emissary (prophet) to the Bajoran people.

SISKO, JAKE (DS9). Human. Teenage son of Benjamin Sisko. Lives aboard Deep Space Nine. Ambition and some talent for writing.

SPOCK (TOS). Vulcan-Human. First officer of the original *Enterprise*, also science officer. Friend and verbal sparring partner of Leonard McCoy. Killed in attempt to save crewmates; later brought back to life. Known for Vulcan pointy ears, green blood, and unrelenting faith in the power of logic.

TORRES, B'ELANNA (VOY). Klingon-Human. Chief engineer of *Voyager*. Formerly a Maquis freedom fighter. After years of denial, embraces her Klingon heritage: "Klingons have a romantic side—it's a bit more vigorous than most."

TROI, DEANNA (TNG). Betazoid-Human. Trained psychologist; ship's counselor, *Enterprise*-D. Her Betazoid origins confer ability to read people's thoughts. Proud of her passion for things containing chocolate; less proud of her flashy, loud, pants-chasing mother, Lwaxana.

TUVOK (VOY). Vulcan. Security officer, *Voyager*. Unrelentingly logical, unwaveringly dispassionate. Takes a dim view of human romance: "Intense romantic love is nothing more than a set of stereotypical behaviors."

UHURA (TOS). Human. Communications officer on the original *Enterprise*.

VORTA (DS9). Humanoid administrators for the Dominion. Overseers of the Jem'Hadar. At least some Vorta have multiple lives—for example, Weyoun.

VOYAGER (VOY). Name given to space probes launched by NASA in early 1970s, and to a United Federation of Planets starship, captained by Kathryn Janeway. On its first mission, the starship *Voyager* is swept up in a massive displacement wave and thrown off course—way off course, into the Delta Quadrant. Under ideal conditions, the trip home would take seventy years—but conditions never seem to be anything close to ideal.

VULCANS (TOS, TNG, DS9, VOY). Homeworld: Vulcan. Vegetarian race that has overcome its violent tendencies by strict adherence to logic and rationality. Vulcan males must mate every seven years or they die. Most distinctive facial feature is pointy ears, a feature shared with Romulans, a closely related species.

WORF (TNG, DS9). Klingon. Chief security officer, *Enterprise*-D. Succeeds to position when Tasha Yar is killed. Poker-playing buddy of Data, LaForge, Troi, and Riker. Later, assigned to Deep Space Nine. Raised by humans following the death of parents. Marries Jadzia Dax despite his family's disapproval. Favorite beverage is prune juice, which he terms the drink of warriors. Courageous and unafraid of possible death, which he faces heroically: "Today is a good day to die."

YAR, TASHA (TNG). Human. Chief security officer, *Enterprise*-D. Has sex with Data to test his claim to anatomical correctness. Killed by malicious alien life-form resembling an oil slick, or the La Brea tar pit in southern California.

Star Trek Episodes Mentioned in This Book

T
he cryptic initials, mysterious digits, and titles that appear throughout this book pinpoint the *Star Trek* episode or movie that we're discussing. An episode from a television series is designated by a three-letter abbreviation of its series name, along with a number that designates the season of its series in which it first aired. The notation "TOS-1," for example, means that the episode in question was aired during the first season of *the original series*. "TNG," "DS9," and "VOY" signify *Star Trek: The Next Generation, Deep Space Nine*, and *Star Trek: Voyager*, respectively. The title given, of course, is that of the episode itself. When we cite one of the *Star Trek* movies, we give its full title. Here, the number in parentheses indicates the chapter(s) in which the episode is mentioned.

This book deals with *Star Trek's* so-called canonical materials. Originally, the word *canonical* was applied to religious books that were certified as authentic or divinely inspired. Today, literary scholars use the word in a different way. "Canonical" no longer implies divine inspiration; instead, it means that a work is central or particularly important. *Star Trek's* canon is most often identified as all the episodes of the four major *Star Trek* television series and the eight feature-length films. The canon excludes material from the animated television series and from the scores of novels based on *Star Trek*. By limiting ourselves to canonical materials, we hope to make it easier for interested readers to find and see for themselves what we're talking about.

These citations will help readers find the sources that spark various discussions in this book. Merely reading about an episode is no substitute for seeing it. All eight feature films, and all of the television episodes except for the most recent, are available on video from Paramount. Whether you've seen a particular episode or film only once or a dozen times, viewing it again is likely to offer some things that you hadn't noticed before. In writing this book, we hope to encourage viewers to see for themselves—for the first time or the tenth.

Star Trek (The Original Series)

TOS-1, "Arena" (4)
TOS-1, "Balance of Terror" (4)
TOS-1, "City on the Edge of Forever" (1)
TOS-1, "Court Martial" (6)
TOS-1, "The Devil in the Dark" (3, 4)
TOS-1, "The Enemy Within" (2, 4)
TOS-1, "The Man Trap" (2)
TOS-1, "The Menagerie" (1)
TOS-1, "The Naked Time" (3)
TOS-1, "Space Seed" (8)
TOS-1, "Taste of Armageddon" (7)
TOS-1, "This Side of Paradise" (2, 3)
TOS-1, "Tomorrow Is Yesterday" (3)

TOS-2, "The Apple" (2)
TOS-2, "Dagger of the Mind" (6)
TOS-2, "The Galileo Seven" (1)
TOS-2, "Patterns of Force" (1)
TOS-2, "Return to Tomorrow" (3)
TOS-2, "The Trouble with Tribbles" (3, 4)

TOS-3, "And the Children Shall Lead" (7)
TOS-3, "Day of the Dove" (2)
TOS-3, "The Empath" (8)
TOS-3, "For the World is Hollow and I Have Touched the Sky" (3)
TOS-3, "Is There in Truth No Beauty?" (5)
TOS-3, "Plato's Stepchildren" (1, 2, 5)
TOS-3, "Requiem for Methuselah" (2)
TOS-3, "The Savage Curtain" (4)
TOS-3, "Spock's Brain" (1, 2)
TOS-3, "The Tholian Web" (4)
TOS-3, "Turnabout Intruder" (3)

Star Trek: The Next Generation

TNG-1, "11001001" (8)
TNG-1, "Angel One" (3, 4)
TNG-1, "Code of Honor" (8
TNG-1, "Conspiracy" (4)
TNG-1, "Encounter at Farpoint," I (4, 8)

TNG-1, "Encounter at Farpoint," II (5)
TNG-1, "Heart of Glory" (5)
TNG-1, "Justice" (4, 6)
TNG-1, "The Naked Now" (3)
TNG-1, "Symbiosis" (3, 7)
TNG-1, "Where No One Has Gone Before" (4)

TNG-2, "Amok Time" (3)
TNG-2, "The Child" (3, 4)
TNG-2, "The Dauphine" (3)
TNG-2, "The Emissary" (3, 4)
TNG-2, "The Icarus Factor" (4)
TNG-2, "Manhunt" (3)
TNG-2, "The Measure of a Man" (6)
TNG-2, "The Outrageous Okona" (2, 4)
TNG-2, "Peak Performance" (6)
TNG-2, "Samaritan Snare" (4)
TNG-2, "Where Silence Has Lease" (8)

TNG-3, "Best of Both Worlds," I (2)
TNG-3, "Booby Trap" (8)
TNG-3, "The Enemy" (5)
TNG-3, "Hollow Pursuits" (7)
TNG-3, "The Hunted" (4)
TNG-3, "A Matter of Perspective" (8)
TNG-3, "Ménage à Troi" (5)
TNG-3, "The Most Toys" (5)
TNG-3, "The Offspring" (2, 3)
TNG-3, "The Price" (8)
TNG-3, "Tin Man" (8)
TNG-3, "Transfiguration" (3, 5, 6)

TNG-4, "Brothers" (2)
TNG-4, "Half a Life" (7)
TNG-4, "The Loss" (8)
TNG-4, "Night Terrors" (7)
TNG-4, "The Nth Degree" (6)
TNG-4, "Redemption," I (4)
TNG-4, "Reunion" (3, 4)
TNG-4, "In Theory" (5)
TNG-4, "The Wounded" (5)

TNG-5, "Conundrum" (6)
TNG-5, "Cost of Living" (8)
TNG-5, "Ethics" (5, 7, 8)
TNG-5, "The Game" (4)
TNG-5, "Hero Worship" (5, 6)
TNG-5, "I, Borg" (3)
TNG-5, "Imaginary Friend" (1)
TNG-5, "The Masterpiece Society" (5)
TNG-5, "New Ground" (3)
TNG-5, "The Outcast" (3)
TNG-5, "Redemption," II (4)
TNG-5, "Silicon Avatar" (1, 3)
TNG-5, "Violations" (6)

TNG-6, "Chain of Command," I (3)
TNG-6, "The Chase" (3)
TNG-6, "Descent," I (2, 8)
TNG-6, "Frame of Mind" (1, 7)
TNG-6, "Lessons" (5)
TNG-6, "Realm of Fear" (7)
TNG-6, "Relics" (7, 8)
TNG-6, "Second Chances" (2, 8)
TNG-6, "Tapestry" (4)
TNG-6, "Unification," II (2)

TNG-7, "All Good Things . . ." (3, 5, 7, 8)
TNG-7, "Attached" (8)
TNG-7, "Descent," II (2)
TNG-7, "Eye of the Beholder" (4)
TNG-7, "Interface" (7)
TNG-7, "Liaisons" (4)
TNG-7, "Lower Decks" (2)
TNG-7, "Parallels" (8)
TNG-7, "Thy Own Self" (8)

Star Trek: Deep Space Nine

DS9-1, "Babel" (3, 7)
DS9-1, "Duet" (7)
DS9-1, "Captive Pursuit" (2)
DS9-1, "Emissary" (4)
DS9-1, "If Wishes Were Horses" (3, 8)
DS9-1, "A Man Alone" (3, 8)

DS9-1, "The Passenger" (2)
DS9-1, "Progress" (4)
DS9-1, "Prophet Motive" (8)
DS9-1, "Q-Less" (3)
DS9-1, "The Storyteller" (1)

DS9-2, "The Alternate" (8)
DS9-2, "Blood Oath" (8)
DS9-2, "The Collaborator" (7)
DS9-2, "Indiscretion" (3)
DS9-2, "Melora" (5)
DS9-2, "The Muse" (3)
DS9-2, "Necessary Evil" (2)
DS9-2, "Paradise" (2)
DS9-2, "Playing God" (5)
DS9-2, "Rules of Acquisition" (6)
DS9-2, "Sanctuary" (4)
DS9-2, "Shadowplay" (3, 8)
DS9-2, "The Wire" (7)

DS9-3, "The Abandoned" (7, 8)
DS9-3, "Alter Ego" (8)
DS9-3, "Equilibrium" (2)
DS9-3, "Heart of Stone" (8)
DS9-3, "The House of Quark" (3)
DS9-3, "Life Support" (7)
DS9-3, "Meridian" (8)
DS9-3, "The Search" (3)
DS9-3, "Shaakar" (4)
DS9-3, "Visionary" (4)

DS9-4, "Body Parts" (8)
DS9-4, "Hard Time" (4, 7)
DS9-4, "Hippocratic Oath" (7, 8)
DS9-4, "Little Green Men" (1, 6)
DS9-4, "Starship Down" (8)
DS9-4, "To The Death" (4)

DS9-5, "Apocalypse" (4)
DS9-5, "By Inferno's Light" (7)
DS9-5, "By Purgatory's Light" (2, 3)
DS9-5, "Children of Time" (4)
DS9-5, "Darkness and the Light" (5)
DS9-5, "Doctor Bashir, I Presume" (6, 8)
DS9-5, "For the Uniform" (5)

DS9-5, "Let He Who Is Without Sin" (3)
DS9-5, "Looking for par'Mach in All the Wrong Places" (3, 5)
DS9-5, "Nor the Battle to the Strong" (1, 2)
DS9-5, "The Ship" (3)
DS9-5, "A Simple Investigation" (3, 4)

DS9-6, "Rocks and Shoals" (7)
DS9-6, "Sons and Daughters" (7)
DS9-6, "A Time to Stand" (4)

Star Trek: Voyager

VOY-1, "Caretaker" (3)
VOY-1, "Parturition" (3)

VOY-2, "Deathwish" (7)
VOY-2, "Elogium" (3)
VOY-2, "False Profits" (6)
VOY-2, "Flashback" (6)
VOY-2, "Projections" (8)

VOY-3, "Alter Ego" (2)
VOY-3, "Blood Fever" (3)
VOY-3, "The Chute" (4)
VOY-3, "Favorite Son" (3)
VOY-3, "Future's End," I (6)
VOY-3, "Remember" (6)
VOY-3, "Rise" (8)
VOY-3, "The Swarm" (5, 6)

VOY-4, "Scorpion," II (2)

Star Trek Movies

Star Trek: The Motion Picture (2, 6, 7, 8)
Star Trek II: The Wrath of Khan (3, 5, 6, 7, 8)
Star Trek III: The Search for Spock (6)
Star Trek IV: The Voyage Home (3, 5)
Star Trek VII: Star Trek Generations (2)
Star Trek VIII: First Contact (3, 4, 5)

Star Trek (The Original Series)—Animated Version

TOS-A, "The Practical Joker" (8)

Notes

1. It appears that pips were not used in Starfleet during the twenty-third century. Uniforms in the original television series do not display them as signs of rank.

2. T. Grandin, *Thinking in Pictures and Other Reports from My Life with Autism* (New York: Doubleday, 1995), 108.

3. Grandin, *Thinking in Pictures*, 113.

4. Grandin, *Thinking in Pictures*, 131–32. Dr. Grandin is probably describing "The Galileo Seven" (TOS-1).

5. In fact, the original *Star Trek* series *was* canceled after its first three years on the air. But the persistent voices of its fans, plus the huge success of the reruns, encouraged network television to resurrect the *Star Trek* concept. The huge success of the seven-season *Star Trek: The Next Generation* in turn spawned the two latest series, *Star Trek: Deep Space Nine* and *Star Trek: Voyager.*

6. S. J. Davies, P. M. Field, and G. Raisman, "Embryonic tissue induces growth of adult axons from myelinated fiber tracts," *Experimental Neurology* 145 (1997): 471–76; W. Roush, "Are pushy axons a key to spinal cord repair?" *Science* 276 (1997): 1971–72.

7. L. M. Krauss, *The Physics of Star Trek* (New York: Basic Books, 1995).

8. I. Asimov, "Social science fiction," in D. Knight (ed.), *Turning Points: Essays on the Art of Science Fiction* (New York: Harper & Row, 1977), 29–61.

9. R. W. Weisberg, *Creativity: Beyond the Myth of Genius* (New York: W. H. Freeman, 1993).

10. R. A. Finke, T. B. Ward, and S. M. Smith, *Creative Cognition: Theory, Research, and Applications* (Cambridge, Mass.: MIT Press, 1992).

11. The professionals who work on *Star Trek* create their aliens just as the rest of us might, cobbling them together out of familiar bits and pieces. Michael Westmore, makeup artist and designer of aliens for *Star Trek: Deep Space Nine* and *Star Trek: Voyager*, gets some of his ideas from *National Geographic* and other magazines. For example, the ridges that grace Klingon foreheads had roots right here on Earth, in dinosaur bones that Westmore happened to see (*The Press-Enterprise*, Riverside, California, 21 September 1997).

12. The use of alternative timelines represents one of *Star Trek's* most popular story line devices. In *Star Trek Chronology: The History of the Future* (New York: Pocket Books, 1996), Michael Okuda and Denise Okuda identify over two dozen episodes from all four *Star Trek* series in which crews are thrown back into the past or catapulted into the future.

13. Aristotle would also have been amazed at his student Plato's enduring universewide influence, particularly on the planet Platonius, whose humanoid inhabitants have set up their own less-than-perfect version of Plato's ideal society (TOS-3, "Plato's Stepchildren").

14. S. Finger, *History of Neuroscience* (Oxford: Oxford University Press, 1994), 10.

15. M. Minsky, *The Society of Mind* (New York: Simon & Schuster, 1986), 287.

16. R. F. Thompson, *The Brain* (New York: W. H. Freeman, 1993), 1.

17. For an excellent discussion of how science fiction stimulates science, see R. A. Heinlein, "Science fiction: Its nature, fault, and virtues," in D. Knight (ed.), *Turning Points: Essays on the Art of Science Fiction* (New York: Harper & Row, 1977), 3–28.

18. The ability of random patterns to instigate the perception of faces was described by F. W. Campbell in a chapter entitled "How much of the information falling on the retina reaches the visual cortex and how much is stored in visual memory?" in C. Chagas, R. Gattass, and C. Gross (eds.), *Pattern Recognition Mechanisms* (Berlin: Springer-Verlag, 1986, 83–95).

➤ ➤ ➤ Chapter 2

1. This highly unusual emotional reaction was made possible because Data's brother, Lore, had managed to alter Data's brain.

2. Quoted by Charles Darwin in *The Expression of the Emotions in Man and Animals* (New York: D. Appleton and Company, 1872/1899).

3. The popularity of e-mail has produced a new instantiation of emotional expressions. Called "emoticons" or "smileys" these are small sequences of keyboard characters that create stylized faces. Some of the more expressive emoticons include:

:)	or : -)	"smile"	: - /	"perplexed"
: - D		"big smile"	: ~(	"sad"
: (	or : -(	"frown"	; -)	"wink"

4. R. W. Levenson, P. Ekman, L. Heider, and W. V. Friesen, "Emotion and autonomic nervous system activity in the Minangkabau of West Sumatra," *Journal of Personality and Social Psychology* 62 (1992): 972–88.

5. We first learn about the emotion chip when Data visits his creator, Dr. Noonien Soong (TNG-4, "Brothers"). Data is shown the tiny microcircuit that Soong designed for him, which will give him the emotions he's never before had. The chip, however, ends up in the positronic brain of Data's identical twin, Lore, who escapes, leaving Data with his emotionless personality. In a later episode (TNG-6, "Descent," I) Lore reappears, with a cadre of Borg under his command. Data is involuntarily drawn into Lore's evil plot to overthrow biological life-forms and replace them with intelligent machines like himself and the Borg. Eventually, Data breaks free of Lore's hold over him, deactivates Lore's positronic brain, and removes the emotion chip. Data is prepared to destroy the chip, but his friend Geordi LaForge urges him to reconsider: "Data, I wouldn't be very much of a friend if I let you give up on a lifelong dream, now would I? Maybe . . . someday, when you're ready . . ." (TNG-7, "Descent," II).

6. Data is not the only character who can be immobilized by fear. Captain Benjamin Sisko's teenage son Jake finds himself in the midst of a Klingon attack on a medical facility (DS9-5, "Nor the Battle to the Strong"). Jake is terrified, and hides in a gully rather than escaping with his companion, Dr. Julian Bashir. The two get separated, and Bashir is injured. Jake's guilt at abandoning Dr. Bashir motivates him to overcome his fear the next time he's confronted with danger—and his clumsy aim with a phaser nonetheless halts a Klingon advance party, allowing medical personnel and patients to escape. Reflecting back on these experiences, Jake realizes that he was just as scared during the second attack as he was during the first; he wasn't, however, immobilized. Jake also comes to appreciate the fine line between courage and cowardice.

7. Her deteriorating condition and violent mood swings are fixed when Dax returns to the Trill homeworld to be reunited with one of the previous hosts of the Dax symbiont.

8. E. Diener and C. Diener, "Most people are happy," *Psychological Science* 7 (1996): 181–85.

9. D. Lykken and A. Tellegen, "Happiness is a stochastic process," *Psychological Science* 7 (1996): 186–89.

10. Results from family studies, including comparisons of twins with one another, are difficult to interpret because it's hard to weigh the relative contributions of environment and heredity. There are several possible nongenetic reasons why close relatives might exhibit similar traits or behaviors. They do, of course, have more genes in common than do unrelated individuals. But in most cases, close blood relatives also live in the same kinds of environments, so their comparable experiences also contribute to their similarities. Studies of fraternal and identical twins can disentangle the contributions of heredity and environment to some extent, since identical twins have identical chromosomal DNA, whereas fraternal twins' chromosomal DNA is no more alike than that of nontwin siblings. To the extent that identical twins are more similar in some trait than are fraternal twins, we should suspect a significant genetic contribution to that trait. This conclusion, however, is based on the questionable assumption that the environmental forces experienced by identical twins are no more similar than those experienced by fraternal twins. A more reliable test is provided by comparisons of identical twins reared apart, ideally in very different environments. Along similar lines, adopted children can be compared with their biological and their adoptive parents.

11. D. H. Hamer, "The heritability of happiness," *Nature Genetics* 14 (1996): 125–26.

12. S. Schama, "Balmorality," *The New Yorker*, 11 August 1997: 38–41.

13. P. Ekman, *Telling Lies: Clues to Deceit in the Marketplace, Politics, and Marriage* (New York: W. W. Norton & Company, 1992).

14. It is interesting to note that whatever made some of the nurses good liars also seems to have helped them in class: the same nurses who managed to mask their true emotions in the experiment were also the ones who got the best grades during the remaining years of their training.

15. This episode has special historical status in the annals of *Star Trek*: it was the first episode to be televised.

16. *Limbic* comes from the Latin *limbus*, meaning "rim" or "border," in recognition of the limbic system's archlike shape.

17. As its name implies, any tricorder, medical or otherwise, comprises three components: sensors that can pick up various forms of energy, a computer that performs sophisticated analyses on the sensors' input, and an all-purpose recorder. A simple handheld tricorder can penetrate stone or metal walls to show what's inside (TOS-1, "The Devil in the Dark"; TOS-2, "The Apple"); analyze the materials used in a painting or the makeup of a life-form (TOS-3, "Requiem for Methuselah"); tell whether someone is telling the truth (TOS-3, "Spock's Brain"), or even tell whether a person is alive or dead (DS9-1, "Emissary"; DS9-1, "The Passenger").

18. For example, if a person is transported from place to place within a starship, even a tiny error could trap the hapless transportee inside a solid object, like a steel wall or computer console (TOS-3, "Day of the Dove"). This risk led to a ban on intraship transporter use.

19. At first, they're impossible to tell apart except by the color difference of their uniforms. After Kirk I's face is deeply scratched by Yeoman Rand, it's easy to tell them apart—at least until Kirk I puts a similar scratch on Kirk II's face.

20. A. R. Damasio, *Descartes' Error* (New York: Grosset, 1994), 44.

21. P. J. Elsinger and A. R. Damasio, "Severe disturbance of higher cognition after bilateral frontal lobe ablation: Patient EVR," *Neurology* 35 (1985): 1731–41.

22. C. A. Weaver III, "Do you need a 'flash' to form a flashbulb memory?" *Journal of Experimental Psychology: General* 122 (1993): 39–46.

➤ ➤ ➤ Chapter 3

1. Twentieth-century computer users, too, are familiar with replication by exploitation. Small bits of software code, also called "viruses," can surreptitiously insert themselves into the operating system of a computer, destroying files and shutting down operations altogether. The viral metaphor is apt: the code on its own contains the instructions for replicating itself, but it can do so only by exploiting the resources of the host computer that it infects.

2. Discussions of evolution often focus on "fitness" and "survival" as the driving forces in the development of new inherited characteristics. From the standpoint of a species' continued existence, however, the ultimate goal must be to pass on one's genetic makeup to offspring who themselves develop into reproductively successful organisms. Survival is important only to the extent that the individual lives long enough to produce offspring and, where necessary, nurture those offspring to their own sexual maturity. For species that rely on sexual reproduction, the cornerstone of evolution is sex, not survival.

3. This species-saving underground city had been created not by the Ocampa, but by more advanced life-forms who had come from another planet.

4. Biologist Richard Dawkins points out that the mutations would run wild within a species if the mutation rate were left unchecked. He points out that natural selection provides the brake on the speed at which mutations spread: "[W]hat we find is that natural selection exerts a braking effect on evolution. The baseline rate of evolution, in the absence of natural selection, is the maximum possible rate. That is synonymous with the mutation rate" [*The Blind Watchmaker* (New York: W. W. Norton, 1986; reissued 1996), 125]. Incidentally, the incidence of harmful mutations within the gene pool would increase if blood relatives freely mated with one another (behavior technically known as consanguineous mating, and colloquially known as inbreeding). Because close relatives have many genes in common, inbreeding increases the chance that dangerous recessive mutations will express themselves. Proscription against sex between close relatives reduces the incidence of deleterious mutations in the gene pool.

5. Class-M planets are Earthlike planets capable of sustaining humanoid life. A key characteristic of these planets is the presence of an oxygen-nitrogen atmosphere.

6. R. E. Leakey and R. Lewin, *Origins* (New York: E. P. Dutton, 1977).

7. Technically speaking, a hominid is any member of the family Hominidae. Members of this family includes us—*Homo sapiens*—as well as earlier humanlike species, including the earliest known bipedal hominid, *Australopithecus afarensis*, whose brain size was about the same as that of a modern-day chimpanzee.

8. As Dawkins puts it, "[T]he mathematical space of all possible trajectories is so vast that the chance of two trajectories ever arriving at the same point becomes vanishingly small" (*The Blind Watchmaker*, 94).

9. Biologically speaking, a *species* is defined as "a population whose members are able to interbreed freely under natural conditions" [E. O. Wilson, *The Diversity of Life* (New York: W. W. Norton, 1996), 38]. So, by this definition, these diverse creatures belong to the same species. And these are just the matings that result in offspring. If we include episodes involving sex (but not necessarily fertilization) between different *Star Trek* creatures, the participants in the mating game expand to include even more aliens: Ferengi and Trills copulate with Klingons (DS9-5, "Looking for par'Mach in All the Wrong Places"), androids have intercourse with humans (TNG-1, "The Naked Now"), and Ocampa and Talaxians are at least aroused by the idea of sex with one another (VOY-2, "Elogium").

10. The blueprint analogy is useful in discussing sexual reproduction, but oversimplifies biochemical realities. In architecture, one blueprint specifies the materials and design for a specific structural component of a building; different blueprints deal with different components. In contrast, the biochemical instructions specifying the "materials" and "design" for a given biological structure may be scattered over multiple genes located on different chromosomes; a single gene carries instructions for making just a single protein. Moreover, the actions (or *expression*, as it's termed) of a given gene must be very carefully programmed into the developmental sequence—a given gene may be able to accomplish its job only when the actions of other genes have been completed.

11. Geneticists estimate that offspring survival is impossible once the total number of genes exceeds the normal number by 4 percent or falls 2 percent below normal. And even smaller deviations are associated with severe physical and mental disorders. The condition known as Down's syndrome is associated with the presence of an extra chromosome number 21. Besides having physical anomalies (e.g., short stature), individuals with this congenital condition are mentally retarded and unusually susceptible to respiratory disease. In contrast, the condition known as Prader-Willi syndrome is associated with the absence of particular genes involved in the production of proteins found in certain parts of the brain—individuals with Prader-Willi syndrome lack those proteins. The resulting symptoms include obesity and mental retardation.

12. There are several other reasons why males and females from different species cannot successfully interbreed. For starters, the male sperm must possess the correct enzyme to allow penetration of the female egg membrane just to gain access to the chromosomes in the egg. Assuming successful egg penetration, the chromosomes carried by the sperm must be able to fuse with those contained in the egg, a process that depends on biochemical compatibility of the two sets of chromosomes. And even if that works, the fertilized egg must successfully attach itself to the uterine wall and establish physical connections with the blood vessels in that wall. An egg fertilized by sperm from another species would most likely be treated as a foreign substance and rejected by antibodies in the mother's body. All things considered, sexual reproduction between genetically compatible mates comes close to being a miracle. Sexual reproduction between genetically diverse mates defies the laws of nature.

Of course, there are examples of successful interspecies breeding. Tigers and lions, for example—two species that never attempt to mate in the natural environment—can produce viable offspring when mated in captivity. The resulting hybrids are termed "tiglons" when the father is a tiger and "ligers" when the father is a lion. A more familiar example of interbreeding involves forced matings between horses and donkeys—the hybrid offspring are mules. These successful hybridizations are possible only because the two species recently evolved from a common ancestor and therefore have a great deal of overlap in their genetic makeup.

13. Several noted scientists have scoffed at *Star Trek's* ignorance of the laws of probability when it comes to genetics and interbreeding. For example, the late Carl Sagan wrote that "the idea that Mr. Spock could be a cross between a human being and a life-form independently evolved on the planet Vulcan is genetically far less probable than a successful cross of a man and an artichoke" [*The Demon-Haunted World* (New York: Ballantine, 1995), 375]. But Sagan may have misunderstood. *Star Trek's* writers did *not* believe that these life-forms evolved independently on different planets. As these episodes demonstrate, the writers were aware of the improbability of such events, and they tried to offer plausible explanations for the ubiquitous humanoids.

14. R. Dawkins, *The Selfish Gene* (New York: Oxford University Press, 1976).

15. One recently evolved, harmless meme was identified by the creator of the meme meme, Richard Dawkins: the wearing of baseball caps backwards. The origin of this meme is unknown, but its success in replicating in the mind of one teenage boy after another is undeniable [R. Dawkins, "Viruses of the mind," in B. Dalhbom (ed.), *Dennett and His Critics: Demystifying Mind* (Cambridge, Mass.: Blackwell, 1993)]. Basketball fans may recognize another, similar meme: the wearing of baggy uniform shorts. For this epidemic, Patient Zero was probably Michael Jordan, who introduced the meme during his early days as a Chicago Bull. This fashion meme was then popularized at the college level by a talented group of University of Michigan basketball players, collectively dubbed "the Fab Five."

16. S. LeVay, *The Sexual Brain* (Cambridge, Mass.: MIT Press, 1993).

17. Actually, Odo's first step in his long overdue sexual initiation *did* take place in Quark's bar. It was there that Odo first encountered Arissa, the Idanian female whose beauty and charm lured him into the bedroom. (DS9-5, "A Simple Investigation").

18. T. Perper, *Sex Signals: The Biology of Love* (Philadelphia: ISI Press, 1985).

19. Perper, *Sex Signals*, 78.

20. H. E. Fisher, *Anatomy of Love: The Natural History of Monogamy, Adultery, and Divorce* (New York: W. W. Norton, 1992).

21. E. T. Rolls, M. K. Sanghera, and A. Roper-Hall, "The latency of activation of neurons in the lateral hypothalamus and substantia innominata during feeding in the monkey," *Brain Research* 164 (1979): 121–35.

22. D. Symons, *The Evolution of Human Sexuality* (New York: Oxford University Press, 1979).

23. This surprising statistic is given by biologist Robin Baker in *Sperm Wars* (New York: Basic Books, 1996). According to him, the precise figures vary with socioeconomic background. To quote Baker (124–25): "On average, about 10 percent of children are not sired by their supposed fathers. Some men, however, have a higher chance of being deceived in this way than others—and it is those of low wealth and status who fare worst. Actual figures range from 1 percent in high-status areas of the United States and Switzerland, to 5 to 6 percent for moderate-status males in the United States and Great Britain, to 10 to 30 percent for lower-status males in the United States, Great Britain, and France. Moreover, the men most likely to sexually hoodwink the lower-status males are men of higher status."

24. B. P. Buunk, A. Angleitner, V. Oubaid, and D. M. Buss, "Sex differences in jealousy in evolutionary and cultural perspective: Tests from the Netherlands, Germany, and the United States," *Psychological Science* 7 (1996): 359–63.

25. D. A. DeSteno and P. Salovey, "Evolutionary origins of sex differences in jealousy? Questioning the 'fitness' model," *Psychological Science* 7 (1996): 367–72.

26. R. Nisbett and T. D. Wilson, "Telling more than we can know: Verbal reports on mental processes," *Psychological Review* 84 (1977): 231–59.

27. Fisher, *Anatomy of Love.*

28. *Star Trek* suggests that sex roles will continue to change, but at a glacial pace. In the twenty-third century, female crew members aboard NCC-1701 play only supporting roles: communications officer (Lieutenant Uhura) and chief nurse (Lieutenant Chapel), with occasional appearances by Captain Kirk's secretary. The important positions—captain, first officer, doctor, chief engineer—go to males (Kirk, Spock, McCoy, Scott). Indeed, there is some reason to believe that women were ineligible to serve as Starfleet captains as late as the mid-twenty-third century (TOS-3, "Turnabout Intruder"), although that had certainly changed by the end of the twenty-third century, when we see a female captain aboard the U.S.S. *Saratoga* in *Star Trek IV: The Voyage Home.*

With the commissioning of NCC-1701D, the complement of females among the officer staff enlarges: Dr. Crusher (replaced for a year by Dr. Pulaski), Counselor Troi, Security Officer Yar (for one year) and Guinan, proprietor of Ten-Forward. Still, all key positions in command, engineering, and in later seasons, security are in the hands of men (Picard, Riker, Worf, Data, LaForge); the nurturing roles are handled by females. Two of the TNG female crew members are also mothers: Crusher for the duration of her duty on the *Enterprise*-D, and Troi for a few days (TNG-2, "The Child").

Deep Space Nine, being a station and not a traveling vessel, offers different job opportunities. The positions of captain (Benjamin Sisko) and engineer (Miles O'Brien) remain in the hands of males, but the second officer is a female (Kira Nerys). Two positions previously held by women aboard the *Enterprise*-D are occupied by males: doctor (Bashir) and lounge proprietor (Quark). The station's constable (Odo, in a position comparable to security officer) is a changeling housed in a male humanoid body, and the science officer (Jadzia Dax) is an asexual symbiont housed in a female humanoid body.

Voyager's crew is a somewhat ragtag group put together when they found themselves stranded 70,000 light years from home. In a real coup for feminism, the captain is a woman (Janeway), as is her chief engineer (B'Elanna Torres) and a key assistant to the engineer (Seven of Nine). Males assume the roles of first officer (Chakotay), security officer (Tuvok), pilot (Paris), communications officer (Kim), and guide/cook (Neelix). The medical assistant is female (Kes), and the doctor is a holodeck program that assumes a male, human appearance.

It is also worth noting that a blow for female equality was struck early in the original series when the computer on board the *Enterprise* was reprogrammed with a female voice. According to Captain Kirk (TOS-1, "Tomorrow is Yesterday"), this change was implemented during a maintenance stop at Cygnet XIV, a society dominated by women. The engineers there felt that the *Enterprise's* computer lacked character and, therefore, gave it a new, female voice. The voice actually used was that of Majel Barrett, who played Nurse Chapel and Lwaxana Troi. Barrett was the wife of *Star Trek* creator Gene Roddenberry.

➤ ➤ ➤ Chapter 4

1. For this and related discoveries, Hess received the Nobel Prize in 1949. Hess's work is summarized in his *Diencephalon: Autonomic and Extrapyramidal Functions* (New York: Grune & Stratton, 1954).

2. Klingon swords, usually wielded with both hands.

3. The game called dom-jot resembles a cross between pool and pinball. Played on an irregularly shaped table, dom-jot is a favorite among the Nausicaans, who don't hesitate to cheat in order to win (TNG-6, "Tapestry").

4. R. A. Baron, *Human Aggression* (New York: Plenum Press, 1977).

5. A. H. Buss and M. Perry, "The aggression questionnaire," *Journal of Personality and Social Psychology* 63 (1992): 452–59.

6. They do not use the word *aggression*, instead using the term *antagonism*, a closely related if not exactly equivalent idea.

7. Domestic violence is a serious health problem in all parts of the world. For example, one-eighth of all homicide cases in the United States involve husbands who murder their wives.

8. L. Berkowitz, *Aggression: Its Causes, Consequences, and Control* (New York: McGraw-Hill, 1993), 264–65.

9. B. Franklin, *Poor Richard's Almanac* (1734).

10. Data's choice of teacher (the part is played by Joe Piscopo) is not an inspired one. His computer-generated teacher's jokes are so lame that it's no wonder Data couldn't get the hang of humor—after all, his teacher doesn't have a particularly firm grasp on the subject himself.

11. One particularly disturbing episode that underscores aggression's dark side is "The Hunted" (TNG-3). An armed, extremely dangerous prisoner named Rogar Dana has broken out of the high-security Angosian penal colony on Lunar V, killing three guards in the process. The *Enterprise*-D answers a plea from the Angosian ambassador for help in recapturing this violent fugitive. When he's beamed aboard the *Enterprise*-D, Dana beats the daylights out of several security personnel, throwing them around like matchsticks and nearly demolishing the transporter room in the process.

Dana may be a killer, but it turns out that he is not of the natural-born variety. Like other Angosians, he was a peaceful person who believed that "reason should settle disputes." But he was altered. When he volunteered for duty in the Tarssian War, Dana underwent intense psychological reorientation and was given a combination of drugs that altered his body's cell structure. Counselor Deanna Troi explains: "He's programmed to be the perfect soldier. He can be absolutely normal, but when a danger is perceived, the program clicks in and takes over. Memory, strength, intelligence, reflexes, all become enhanced. He's conditioned to survive at any cost."

This acquired talent for violence proved invaluable while the Angosians were engaged in the Tarssian wars. But those wars are over, and a now peaceful society has no need for the soldiers it created. They're too dangerous. That's why Dana and his fellow super-soldiers have been locked up on Lunar V. The Angosians face a dilemma like the one that faces all humanoids: Aggression can be useful—in the right time and place—but it can also wreak havoc if allowed to run unchecked.

12. R. E. Leakey and R. Lewin, *People of the Lake* (Garden City, N.Y.: Anchor Press/Doubleday, 1978).

13. The idea that territory and resources form the roots of humanoid aggression finds plenty of support in *Star Trek*. Territory is certainly the root of the conflict between the Cardassians and the Bajorans. And the Bajorans, even after recovering their homeland, fight among themselves over resources (DS9-1, "Progress"; DS9-3, "Shaakar") and block an attempt by other refugee groups to seek sanctuary in an uninhabited region of Bajor (DS9-2, "Sanctuary"). In both TOS and TNG episodes, we hear about an alien race—the Tholians—who are downright xenophobic in their hatred of territorial intruders. In the only episode in which they actually appear, the Tholians use a fascinating maneuver to capture intruders: a pair of Tholian ships sets up interconnecting energy fields that ensnare intruders much like a spiderweb (TOS-3, "The Tholian Web"). One could even argue that the Borg collective is driven by a territorial imperative. But in the case of the Borg, the territory encompasses the individual minds of all non-Borg entities.

14. The Klingon body evolved to accommodate a propensity for battle. Nearly all Klingon vital organs are duplicated, providing insurance against lethal injury [M. Okuda, D. Okuda, and D. Mirek, *The Star Trek Encyclopedia* (New York: Pocket Books, 1994)].

15. J. L. Cloudsley-Thompson, *Animal Conflict and Adaptation* (London: G. T. Foules & Company, 1965).

16. Among those assigned to "the chair" by Picard is Lieutenant Worf (TNG-2, "The Emissary"). Following Worf's turn in the seat, Commander Riker asks him, "How did it feel to be in command?" to which Worf replies, "Comfortable chair." To this terse answer, Worf's paramour K'Ehleyr adds, "And you wore it well."

17. Any human who has had the traumatic experience of transferring from one school to another can sympathize with Riker. Classmates often force a newcomer to pass initiation tests before being admitted to the circle of friends. In college sororities and fraternities, these tests have become institutionalized as "rush week."

18. E. O. Wilson, *On Human Nature* (Cambridge, Mass.: Harvard University Press, 1978).

19. K. Lorenz, *On Aggression* (New York: Harcourt, Brace & World, 1966).

20. Lorenz, *On Aggression*.

21. Here's how anbo-jytsu works. Outfitted with protective armor, each contestant uses a long staff to swing at his opponent, the aim being to score a hit and, if possible, knock his opponent out of the anbo-jytsu ring. Like twentieth-century martial arts, anbo-jytsu isn't intended to provoke feelings of aggression. Usually participants treat it as a test of coordination and speed, with no desire to injure an opponent.

22. An exception to this rule is the inhabitants of the planet Angel One, which was visited by the crew of the *Enterprise*-D (TNG-1, "Angel One"). Here the females, who hold all positions of responsibility, are tall and strike commanding postures. The males, who assume subsidiary roles, are smaller and subservient.

23. Sexually related behaviors in adulthood are related to hormone exposure during fetal infancy. To illustrate, consider what happens to females exposed to masculinizing hormones during their fetal development. First, these females are much more likely to prefer women as sexual partners, whereas women unexposed to these particular hormones overwhelmingly prefer male partners. Second, these "masculinized" females tend to be more aggressive later in life. And third, they tend to perform better on tests of visual-spatial ability, mirroring the differences seen among normally developing males and females [E. B. Maccoby and C. N. Jacklin, *The*

Psychology of Sex Differences (Stanford: Stanford University Press, 1974); M. Hines and R. Green, "Human hormonal and neural correlates of sex-typed behaviors," *Review of Psychiatry* 10 (1991): 536–55]. Males, too, are susceptible to hormonal disturbances during fetal development. If exposed to abnormally low levels of the male sex hormones prior to birth, males exhibit lower than average visual-spatial abilities when tested years later.

24. A. Moir and D. Jessel, *Brain Sex* (New York: Delta Books, 1989), 79.

25. Berkowitz, *Aggression*, 396.

26. The medical literature is full of examples of aggression in humans caused by brain dysfunction [see, for example, J. M. Silver and S. C. Yodofsky, "Violence and the brain," in T. Feinberg and M. J. Farah (eds.), *Behavioral Neurology and Neuropsychology* (New York: McGraw-Hill, 1997), 711–17]. Limbic system tumors frequently are accompanied by heightened irritability, unprovoked rage reactions, and homicidal attacks. Probably the most familiar case is Charles Whitman, the twenty-five-year-old man who murdered his wife and mother one evening and, the next day, climbed to the top of a tower on the campus of the University of Texas, Austin, and used a high-powered rifle to kill fourteen people and wound another twenty-four. An autopsy revealed a tumor growing on one side of his brain in the temporal lobe, a portion of the brain that includes the limbic system. Medical literature documents other cases of increased hostility and irritability in people with tumors in the limbic system, frontal lobes, or hypothalamus. When detected, this condition can be reversed by surgery. The tumor may have its effect by inactivating brain cells that normally inhibit aggressive behavior.

Rabies (a word derived from Latin meaning "rage") is a disease that can cause violent outbursts and irrational assaults. It is caused by transmission of a saliva-borne virus from an infected host (often a dog) to the victim. This virus infects the peripheral nerves and then travels through the spinal cord to the brain. There it spreads throughout the entire brain, in particular affecting brain cells in the limbic system. This specificity raises the possibility that some deviant person could develop a chemical agent that would mimic the effects of the virus, causing involuntary rage reactions in the drugged individual.

27. C. M. Steele and R. A. Josephs, "Alcohol myopia: Its prized and dangerous effects," *American Psychologist* 45 (1990): 921–33.

28. Barroom brawls are rare, but not nonexistent. One well-known incident takes place on Space Station K-7, when some Klingons disparage the U.S.S. *Enterprise* (TOS-2, "The Trouble with Tribbles") and find themselves having to back up their words. As mentioned earlier in this chapter, the Bonestell Recreational Facility—a dingy bar on Starbase Earhart—is the scene of a bloody fight between some Starfleet cadets and a group of Nausicaans (TNG-6, "Tapestry"). In Quark's bar on Deep Space Nine, Worf tears into a Jem'Hadar warrior who has insulted his friend Miles O'Brien (DS9-4, "To The Death"). Quark's bar is also the site for an unusual brawl featuring some drunken Klingons (DS9-3, "Visionary"). Thanks to the magic of time travel, this brawl takes place over and over and over in not-quite-instant replay.

29. In desperate cases, removal of brain tissue can reduce the incidence of uncontrollable aggressive behaviors. Here's an excerpt from the description of one such case involving removal of brain tissue in the temporal lobes: "Prior to the operation the patient had frequent attacks of aggressive and violent behavior during which he had attempted to strangle his mother and to crush his younger brother under his feet. After unilateral temporal lobectomy, he attacked the nurses and doctors and threatened some with death. After the second temporal lobe was removed, he became extremely meek with everyone and was absolutely resistant to any attempt to arouse aggressiveness and violent reactions in him" [K. E. Moyer, *The Psychobiology of Aggression* (New York: Harper & Row, 1976), 46].

30. Wilson, *On Human Nature*, 116.

> > > # Chapter 5

1. We humans have inherited our sensory abilities from our primate ancestors—their habits and needs determined the design of their sensory systems and, therefore, of ours. For example, natural selection deemed it important for our ancestors to have good color vision and excellent binocular depth perception—these visual abilities allowed them to hunt and gather food and to avoid predators. In contrast, it was not so important for our primate ancestors to possess excellent dim-light vision, because they were inactive at night. Consequently, we have trouble seeing at night, unless we flood the visual environment with artificial light. Cats, in comparison, have excellent dim-light vision, inherited from their feline ancestors, whose eyes were designed for hunting at night.

2. Weapons that explode outside the ship, in the vacuum of space, are silent. As Robert Boyle proved long before *Star Trek*, sound cannot propagate in a vacuum, and that's what space is. Explosions on a planet's surface, and explosions generated during combat aboard the *Enterprise*, or the *Defiant*, or *Voyager*, are different matters. There, ears can be traumatized.

3. R. Reagan, *An American Life* (New York: Simon & Schuster, 1990).

4. G. H. Recanzone, C. E. Schreiner, and M. M. Merzenich, "Plasticity in the frequency representation of primary auditory cortex following discrimination training in adult owl monkeys," *Journal of Neuroscience* 13 (1993): 87–103.

5. T. Elbert, C. Pantev, C. Wienbruch, B. Rockstroh, and E. Taub, "Increased cortical representation of the fingers of the left hand in string players," *Science* 27 (1995): 305–7.

6. In this episode, Dr. Jones is played by Diana Muldaur, the same actress who later played the role of Dr. Kate Pulaski in the second season of *Star Trek: The Next Generation*.

7. The name "Medusa" is a tipoff that we're probably not talking about beauty pageant winners. In ancient Terran mythology, Medusa was a creature for whom every day was a bad hair day: she sported a hideous headful of writhing serpents. The sight of Medusa was so terrible that anyone who looked upon her was turned instantly to stone.

8. "IDIC" stands for Gene Roddenberry's *Star Trek* ideal, Infinite Diversity in Infinite Combination.

9. It's easy to tell where this region is in your brain. Just place your right hand horizontally on the back of your skull, with your thumb at the base of your skull. The visual cortex lies directly beneath your hand, at a depth of about 2 cm below your scalp.

10. G. S. Brindley and W. S. Lewin, "The sensations produced by electrical stimulation of the visual cortex," *Journal of Physiology* 196 (1968): 479–93; S. S. Stensaas, D. K. Eddington, and W. H. Dobelle, "The topography and variability of the primary visual cortex in man," *Journal of Neurosurgery* 40 (1974): 747–55. Such experiments are typically carried out in the context of surgical procedures aimed at reducing intractable pain or in preparation for removal of diseased areas of the brain.

11. One particular aspect of electrical brain stimulation deserves underscoring: activation of brain cells within the visual cortex *inside* the skull causes visual sensations that are located in the external world, *outside* the skull. Vision is designed to put us in contact with objects and events in the world, and the neural machinery responsible for vision "projects" the product of its activity out into that world.

12. Also different is the route that the VISOR's information takes. In normal human vision, most signals from the eyes go to a brain structure called the lateral geniculate nucleus, where they are reshuffled before going on to the occipital cortex. This reshuffling lays the groundwork for normal binocular vision. Signals from the VISOR, however, bypass this intermediate stage.

13. R. M. Anderson, "Visual perceptions and observations of an aphakic surgeon," *Perceptual & Motor Skills* 57 (1962): 1211–18.

14. The VISOR performs some data compression before delivering its output. But, based on LaForge's description, the compression is trivial, merely shortening each signal sent to the brain (TNG-5, "The Masterpiece Society").

15. C. Koch and B. Mathur, "Neuromorphic vision chips," *IEEE Spectrum* 33 (1996): 38–46.

16. Sometime after leaving the *Enterprise*-D, Geordi LaForge receives surgically implanted eyes that allow him to discard the VISOR. We see the new, VISOR-less LaForge in "All Good Things . . ." (TNG-7) and in *Star Trek VIII: First Contact*. In twentieth-century medicine there are a few cases in which previously blind people have had their sight restored by eye surgery. In all cases, the surgery involved the removal of dense cataracts present from birth. Thus the brains of these patients had never received patterned visual information. One account of a patient's recovery of sight suggests that due to this lack of previous visual experience, normal vision was never attained (O. Sacks, "To see and not to see," *The New Yorker*, 10 May 1993). In LaForge's case, his brain had been receiving patterned visual input from the VISOR. Because the visual parts of his brain had not been deprived, his postsurgical prognosis is good. It would be fascinating to hear LaForge compare his visual experiences before and after the implants.

17. P. L. Kilbride and H. W. Leibowitz, "Factors affecting the magnitude of the Ponzo perspective illusion among the Baganda," *Perception and Psychophysics* 17 (1975): 543–48.

18. The quantities of various elements in Data's body are enumerated in "The Most Toys" (TNG-3). In that episode, android-nappers make it appear that Data has been destroyed in a shuttlecraft explosion, planting appropriate materials to simulate his remains.

> > > Chapter 6

1. The word amnesia comes from the Greek word meaning "oblivion." When memory for past events is lost, amnesia is termed retrograde (working backward in time). When new memories cannot be formed, amnesia is termed anterograde (working forward in time).

2. Health caregivers often use the term "dementia of the Alzheimer's type" (DAT) to refer to the condition of patients believed to have Alzheimer's disease (AD). A confirmed diagnosis of AD cannot be made until the patient dies and brain tissue can be autopsied. So the diagnosis is an educated guess so long as the patient is alive. But the symptoms are quite reliable—the increasing dementia in particular. Incidentally, although the word demented is colloquially used to mean something like "crazy," medically speaking, the term dementia describes a condition in which the cognitive abilities of the mind (mens in Latin, hence -mentia) are sufficiently compromised so as to interfere with the ability to carry out daily activities independently.

3. "John Doe" (TNG-3, "Transfiguration")

4. P. J. Hilts, *Memory's Ghost* (New York: Simon and Schuster, 1995), 21.

5. The time course of AD varies from one victim to another. Particular cognitive functions are affected earlier or more severely in some people than in others. Despite these variations, there does seem to be a typical progression [S. E. Arnold, B. T. Hyman, J. Flory, A. R. Damasio, and G. W. van Hoesen, "The topographical and neuroanatomical distribution of neurofibrillary tangles and neuritic plaques in the cerebral cortex of patients with Alzheimer's Disease," *Cerebral Cortex* 1 (1991): 103–16].

6. P. J. Eslinger and A. R. Damasio, "Preserved motor learning in Alzheimer's disease: Implications for anatomy and behavior," *Journal of Neuroscience* 6 (1986): 3006–9.

7. There is no consensus on these names. Some researchers prefer the term *explicit memory* in place of *declarative memory* and substitute *habit memory* for *procedural memory* [L. R. Squire and B. J. Knowlton, "Memory, hippocampus, and brain systems," in M. Gazzaniga (ed.), *The Cognitive Neurosciences* (Cambridge: MIT Press, 1994), 825–37]. Whatever the terminology, most researchers believe in the existence of multiple memory systems.

8. AD is not the only brain disorder that reveals the selective roles played by declarative and procedural memory. Another disease that can adversely affect memory is chronic alcoholism, which, among other things, depletes the body of B-complex vitamins. This vitamin depletion causes neural degeneration, resulting in Korsakoff's syndrome, a disorder characterized by impaired memory, problem-solving difficulties, and the propensity to fabricate stories to cover up gaps in memory. This syndrome is named after Sergei Korsakoff, the Russian psychiatrist who first described it around the turn of the twentieth century.

9. The term *echoic memory* is sometimes used to refer to the remembering of sounds for a short period of time. Thus, for example, if you're told by a bank teller that your checking account balance is "five hundred thirty-seven dollars and sixty cents," you hold that auditory information in short-term, echoic memory until you make the entry in your checkbook.

10. C. B. Cave and L. R. Squire, "Intact verbal and nonverbal short-term memory following damage to the human hippocampus," *Hippocampus* 2 (1992): 151–64.

11. The CREB research described here was done in fruit flies [J. C. P. Yin and T. Tully, "CREB and the formation of long-term memory," *Current Opinion in Neurobiology* 6 (1996): 264–68] and in mice [J. H. Kogan, P. W. Frankland, J. A. Blendy, J. Coblentz, Z. Marowitz, G. Schutz, and A. J. Silva, "Spaced training induces normal long-term memory in CREB mutant mice," *Current Biology* 7 (1997): 1–11]. The presence of CREB genes in mammals gives every reason to believe that the same mechanisms contribute to long-term memory in humans.

12. Yin and Tully, "CREB and the formation of long-term memory."

13. About 600 years before Lieutenant Commander Data graduated from Starfleet Academy, the French philosopher Claude-Adrien Helvétius warned his readers about the consequences of putting too much stuff into their memories. Helvétius believed that subpar intelligence was the result of poor choice of memories. Greatness, he claimed, required the courage to remain ignorant of useless things. According to Helvétius, ignorance of unimportant matters is an absolute prerequisite for greatness. Keep that in mind the next time you forget where you put your keys.

14. A. R. Luria, *The Mind of a Mnemonist* (Cambridge, Mass.: Harvard University Press, 1968), 64–65.

15. This memory-boosting strategy is described by I. M. L. Hunter in "Mnemonics," in R. L. Gregory (ed.), *The Oxford Companion to the Mind* (Oxford: Oxford University Press, 1987), 493–94.

16. W. James, *The Principles of Psychology* (Cambridge, Mass.: Harvard University Press, 1981), 628.

17. Working memory and short-term memory share some features, but the two are distinct. Both can be likened to a "scratch pad" on which a limited amount of information is inscribed for immediate use. But working memory's contents are drawn from the permanent records in long-term memory; short-term memory's contents come from the sensory experiences bombarding us all the time. Only a small fraction of short-term memory's contents ever make it to long-term memory. When you are told a new phone number, that information enters short-term memory, where it is available for immediate use. When you recall a familiar phone number, that information enters working memory, where it too is available for use. The two forms of memory are executed by different brain mechanisms. [A. Baddeley, "The fractionation of working memory," *Proceedings of the National Academy of Science* 93 (1996): 13468–72.]

18. J. D. Cohen, W. M. Perlstein, T. S. Braver, L. E. Nystrom, D. C. Noll, J. Jonides, and E. E. Smith, "Temporal dynamics of brain activation during a working memory task," *Nature* 386 (1997): 605–8; S. M. Courtney, L. G. Ungerleider, K. Keil, and J. V. Haxby, "Transient and sustained activity in a distributed neural system for human working memory," *Nature* 386 (1997): 608–11.

19. L. Hasher and R. T. Zacks, "Working memory, comprehension, and aging: A new view," in G. H. Bower (ed.), *The Psychology of Learning and Motivation*, vol. 22 (San Diego, Calif.: Academic Press, 1988), 193–225.

20. These examples are taken from T. Y. Arbuckle and D. P. Gold, "Aging, inhibition, and verbosity," *Journal of Gerontology: Psychological Sciences*, 48 (1993): 225–32.

21. These calculations assume that each floppy is about 1/8 inch thick. These disc equivalents were suggested by Phillip Thorne in a posting to the Internet usergroup rec.arts.startrek.tech in September 1996.

22. A rough estimate of the information generated by human eyes goes this way. There are two eyes, each with about 1.25 million optic nerve fibers delivering information to the brain. Assume that 100 times each second, the information in each nerve fiber is sampled to a precision of 8 bits (1 byte). If this process went on every second of every day for 70 years, the net product of vision alone would be about 4.4×10^{18} bits, more than five times the capacity of Data's brain. Even assuming that sleep allows the system to take a break, the total cumulative visual information would still overflow Data's brain.

23. One lossy scheme, JPEG, is used to compress data from single photographs. Another, called MPEG, deals with sequences of moving images.

24. D. L. Schacter, W. Koutstaal, and K. A. Norman, "Can cognitive neuroscience illuminate the nature of traumatic childhood memories?" *Current Opinion in Neurobiology* 6 (1996): 207–14.

25. E. Tulving, "Memory and consciousness," *Canadian Psychologist* 26 (1985): 1–12.

26. Schacter, Koutstaal, and Norman, "Can cognitive neuroscience illuminate the nature of traumatic childhood memories?"

27. M. Moscovitch, "Confabulation and the frontal systems: Strategic versus associative retrieval in neuropsychological theories of memory," in H. L. Roediger III and F. I. M. Craik (eds.), *Varieties of Memory and Consciousness: Essays in Honour of Endel Tulving* (Hillsdale, N.J.: Lawrence Erlbaum Associates, 1993), 133–60.

28. E. F. Loftus, "Creating false memories," *Scientific American* 277 (September 1997), 70–75.

29. M. Posner, "Seeing the mind," *Science* 262 (1993): 673–74.

30. M. K. Johnson, S. Hashtroudi, and D. S. Lindsay, "Source monitoring," *Psychological Bulletin* 114 (1993): 3–28.

31. H. L. Roediger III and K. B. McDermott, "Creating false memories: Remembering words not presented in lists," *Journal of experimental Psychology: Learning, Memory and Cognition* 21 (1995): 803–14.

32. Lists are taken from T. Beardsley, "As time goes by . . . " *Scientific American* (May 1997): 24.

33. I. E. Hyman Jr. and J. Pentland, "The role of mental imagery in the creation of false childhood memories," *Journal of Memory and Language* 35 (1996): 101–17.

34. "Theta waves" probably refers to 4–7 Hz synchronized, rhythmic activity in many neurons. Theta waves are commonly seen in some sleep states.

35. As we have seen, memory is a collection of abilities, each mediated by a distinct neural system. Dr. Crusher's identification of the thalamus as the "memory center" could refer to the brain region implicated in Korsakoff's syndrome, which damages memory (see note 8). At another time, Dr. Crusher describes the hippocampus as a region of the brain involved in memory (TNG-5, "Conundrum"). This, too, squares with contemporary work on the neurology of memory—the amnesic patient H. M. suffered damage to his hippocampus.

36. D. L. Schacter, E. Reiman, T. Curran, L. S. Yun, D. Bandy, K. B. McDermott, and H. L. Roediger III, "Neuroanatomical correlates of veridical and illusory recognition memory: Evidence from positron emission tomography," *Neuron* 17 (1996): 267–74.

37. J. W. Hall, R. Sekuler, and W. Cushman, "Effects of IAR occurrence during learning on response time during subsequent recognition," *Journal of Experimental Psychology* 79 (1969): 39–42.

➤ ➤ ➤ Chapter 7

1. As a result of the blows he took to the head, Muhammad Ali developed Parkinson's disease. How many of Ali's neurons had to be damaged in order to produce his transformation? Our best guess is about 100,000—which may seem like a lot until you realize that they make up less than a thousandth of one percent of the brain's entire population of neurons. For an excellent account of Ali's brilliant career and a good explanation of how repeated blows to the head caused his disordered condition, see H. L. Kawans, *Why Michael Couldn't Hit* (New York: W. H. Freeman, 1996), 117–47.

2. R. P. Bentall, "The illusion of reality: A review and integration of psychological research on hallucinations," *Psychological Bulletin* 107 (1990): 82–95.

3. D. Chambers and D. Reisberg, "What an image depicts depends on what an image means," *Cognitive Psychology* 24 (1992): 145–74.

4. S. Kosslyn, N. Alpert, W. Thompson, V. Maljkovic, S. Weise, C. Chabris, S. Hamilton, S. Rauch, and F. Buonanno, "Visual mental imagery activates topographically organized visual cortex: PET investigations," *Journal of Cognitive Neuroscience* 5 (1993): 263–87.

5. C. W. Perky, "An experimental study of imagination," *American Journal of Psychology* 21 (1910): 422–52.

6. Eugen Bleuler's coinage of the term *schizophrenia* is described by N. C. Adreasen in "Linking mind and brain in the study of mental illnesses: A project for a scientific psychopathology," *Science* 275 (1997): 1586–93.

7. Remarkably, the killer was judged sane and convicted of murder. He's now serving a prison term of more than 300 years.

8. P. K. McGuire, D. A. Silbersweig, I. Wright, R. M. Murray, R. S. Frackowiak, and C. D. Frith, "The neural correlates of inner speech and auditory verbal imagery in schizophrenia: Relationship to auditory verbal hallucinations," *British Journal of Psychiatry* 169 (1996): 148–59.

9. D. A. Silbersweig, E. Stern, C. Frith, C. Cahill, A. Holmes, S. Grootoonk, J. Seaward, P. McKenna, S. E. Chua, L. Schnorr, T. Jones, and R. S. J. Frackowiak, "A functional neuroanatomy of hallucinations in schizophrenia," *Nature* 378 (1995): 176–79.

10. C. Frith, "The role of the prefrontal cortex in self-consciousness: The case of auditory hallucinations," *Philosophical Transactions of the Royal Society of London*, Series B, 351 (1996): 1505–12.

11. Paul Broca is always given credit for first describing the condition known as aphasia. But, in fact, the honor should go to a little-known

French physician named Marc Dax (no relation, we believe, to Deep Space Nine's Jadzia Dax). Dr. Dax documented a number of cases of aphasia in his patients decades before Broca's discovery. But because he was a practitioner, not a researcher, Dax's obscure presentation at a local medical meeting in 1837 attracted no attention. Yet he described a fascinating association between paralysis on the right side of the body and an inability to speak. His patients, we now know, were suffering left hemisphere brain damage. It remained for Paul Broca, in 1861, to describe his well-publicized patient, whose symptoms exactly mimicked those of Dax's patients.

12. Among the 10 percent of the population who are naturally left-handed, most would develop aphasia only after damage to the left hemisphere; about a quarter of left-handed aphasics develop the condition due to right-hemisphere damage.

13. In the first feature-length *Star Trek* movie—*Star Trek: The Motion Picture*—we actually witness a gruesome transporter accident. Two Starfleet officers attempt to beam from a space shuttle to the *Enterprise*, but the pattern lock on the transporter beam is lost. The two unfortunate officers partially materialize on the *Enterprise's* transporter pad, but then disintegrate in front of our eyes, moaning in agony. Not a pretty sight.

14. Reg Barclay is a smaller, twenty-fourth-century version of John Madden, the burly American football commentator and one-time coach. Madden's twentieth-century job forces him to travel tens of thousands of miles each year—and he does every one of them by bus. Although you can debate the statistics on bus accidents versus airplane accidents, Madden's fear of flying is understandable: airplane accidents do happen, and they can be fatal.

15. People who suffer panic disorders can have episodes even in the absence of external events associated with anxiety. Bouts of panic can be generated entirely internally.

16. Among Americans, this figure is estimated to be more than 10 percent (R. C. Kessler, K. A. McGonagle, S. Zhao, C. B. Nelson, M. Hughes, S. Eshelman, H. U. Wittchen, and K. S. Kendler, "Lifetime and 12-month prevalence of DSM-III-R psychiatric disorders in the United States: Results from the National Comorbidity Study," *Archives of General Psychiatry* 51 (1994): 8–19.

17. M. B. Stein, J. R. Walker, and D. R. Forde, "Public speaking fears in a community sample: Prevalence, impact on functioning, and diagnostic classification," *Archives of General Psychiatry* 53 (1996): 169–74.

18. H. Markus and S. Kitayama, "Culture and the self: Implications for cognition, emotion, and motivation," *Psychological Review* 98 (1991): 224–53.

19. H. C. Triandis, *Culture and Social Behavior* (New York: McGraw-Hill, 1994).

20. One well-known victim of Taijin-Kyofusho was Natsume Soseki, the nineteenth-century Japanese author of *Bottchan* ("A Boy") and

Wagahi-wa ("I'm a Cat"). Soskei, who is considered the Charles Dickens of Japan, developed the disorder while he was studying in England, possibly because he lived in constant fear of violating the unfamiliar norms of English society.

21. R. A. Kleinknecht, D. L. Dinnel, E. Kleinknecht, and N. Hiruma, "Cultural factors in social anxiety: A comparison of social phobia symptoms and Taijin Kyofusho," *Journal of Anxiety Disorders* 11 (1997): 157–77.

22. There are actually two different types of panic attacks: cued and uncued. We concentrate here on so-called uncued attacks, which occur without any obvious trigger.

24. M. Fredrikson, G. Wik, P. Annas, K. Ericson, and S. Stone-Elander, "Functional neuroanatomy of visually elicited simple phobic fear: Additional data and theoretical analysis," *Psychophysiology* 32 (1995): 43–48.

25. Consistent with this idea, tests of recognition or recall reveal that phobic individuals have poorer than normal recall of previously experienced phobic stimuli. These findings have led to the idea that the associated emotional memories are stored and accessed through those pathways involved in implicit memory—the form of memory outside the conscious control of the individual.

26. D. L. Rosenhahn, "On being sane in insane places," *Science* 179 (1973): 250–58.

27. This study has been vigorously criticized on a number of grounds. For example, it's not surprising that these pseudo-patients were admitted to the hospital. The psychiatric staff took the pseudo-patients' admitting symptoms as honest statements of their psychological distress and honored their implicit requests for help. For a good discussion of other criticisms, see R. L. Spitzer, "On pseudoscience in science, logic in remission, and psychiatric diagnosis: A critique of Rosenhahn's 'On being sane in insane places,'" *Journal of Abnormal Psychology* 84 (1975): 442–52.

28. The Obsidian Order implanted this device in Garak's brain to increase his ability to withstand torture if he were captured by their enemies. The pleasurable feelings that would accompany activation of the device would, they reasoned, counteract the pain of torture.

29. Endorphins are members of a family of neurotransmitters manufactured by our brains to alleviate pain and induce a sense of euphoria. Called *opioids*, these neurochemicals are nature's built-in analgesic and reward substances. We don't know much about how opioids produce their effects, but we do know that highly addictive synthetic drugs such as morphine and heroin produce similar effects.

According to the *Enterprise*-D's counselor, Deanna Troi, endorphins have another action besides relieving pain and creating a pleasurable high: they relieve anxiety. Troi teaches Reginald Barclay a self-stimulation technique—called plexing—that is supposed to release endorphins into his brain, thereby relieving the symptoms of his transporter phobia (TNG-6, "Realm of Fear"). Plexing is simple and requires no equipment.

Just take your index and middle fingers and tap lightly with a steady rhythm just behind one of your ears. Exactly why this form of self-stimulation would work—if indeed it does—remains a mystery, although temporary elevation of endorphin levels certainly would act to relieve anxiety.

30. J. Chiles and K. Strosahl, *The Suicidal Patient: Principles of Assessment, Treatment, and Case Management* (Washington, D.C.: American Psychiatric Press, 1995). The incidence of successful suicide varies widely around the world. The highest suicide rates are found among Eastern European countries, including Russia, Estonia, and Hungary—in these countries, the annual incidence of suicide and self-inflicted injury exceeds 0.03 percent (30 per 100,000 people). At the other end of the scale, Central and South American countries tally some of the lowest rates—Argentina, Brazil, and Mexico all have rates of less than 0.003 percent (3 per 100,000 people). The United States, Germany, and Japan are clustered together in the middle of the pack, with suicide/self-injury rates in the 0.014 percent neighborhood. France is a little higher than this (0.02 percent), while England, Spain, and Italy are lower (0.008 percent). Most of these data come from the World Health Organization's database, which can be accessed on the World Wide Web at <<http://www.who.dk/www.who.dk/mainframe.htm>>. This site is chock-full of intriguing facts such as the incidence of missing teeth in Country X, or the number of television sets per 1,000 people in Country Y.

31. This odd contagion of suicide is not a recent invention. In late-eighteenth-century Europe, a number of people were inspired to do themselves in by the dramatic suicide of Werther, the lovelorn title character of Goethe's novel *The Sorrows of Young Werther*.

32. D. Phillips, "The influence of suggestion on suicide: Substantive and theoretical implications," *American Sociological Review* 39 (1974): 340–54.

33. The roster of *Star Trek* characters who contemplate or actually commit suicide includes:

• Starfleet Captain Clark Terrell, who mortally wounds himself with a phaser rather than follow the orders of Khan to murder Captain James Kirk (*Star Trek II: The Wrath of Khan*)

• Amon Marritza, a guilt-ridden Cardassian who wants to kill himself for his role in the Bajoran occupation (DS9-1, "Duet")

• Prylar Bek, a Bajoran monk who hangs himself on the promenade of Deep Space Nine out of guilt over disclosing to the Cardassians the location of forty-two freedom fighters, who were subsequently tracked down and slaughtered (DS9-2, "The Collaborator")

• Scientist-colonists on Triacus, who commit suicide out of fear of their own evil tendencies (TOS-3, "And the Children Shall Lead")

• Inhabitants of Emmeniar VII, who report to suicide camps to fulfill the projected body count from a computer-simulated war with neighboring planet Vendikar (TOS-1, "Taste of Armageddon")

• Inhabitants of Kaelon II, who are required by law to take their own lives when they reach their sixtieth birthday—no ifs, ands, or buts (TNG-4, "Half a Life")

• Alexander Roshenko (Worf's son), who intentionally locks himself in a chamber filled with toxic fumes to end his disgrace at failing in the warrior business (DS9-6, "Sons and Daughters).

34. S. Hollon and A. T. Beck, "Cognitive and cognitive-behavioral therapies," in A. E. Bergin and S. L. Garfield (eds.), *Handbook of Psychotherapy and Behavior Change* (New York: Wiley, 1994), 428–66.

35. While all neuroscientists agree that dopamine plays a rewarding role in the brain, some are skeptical whether this neurotransmitter produces pleasurable feelings *directly*. According to these skeptics, dopamine release signals the presence of a desirable object or a surprising, potentially significant event [I. Wickelgren, "Getting the brain's attention," *Science* 278 (1997): 35–37.] Under this scenario, dopamine's role continues to be the encouragement of adaptive behavior, but rather than acting as a pleasure substance, it operates like a spotlight that focuses attention on environmental events associated with pleasure.

36. Why do some people find it virtually impossible to quit smoking? It may be that hard-core smokers are unwittingly medicating themselves to minimize depression. When they attempt to quit smoking, their depressive symptoms are heightened and they're driven to resume their "medication." This theory makes some sense. The active ingredient in tobacco is nicotine, and nicotine is a drug that stimulates certain brain receptors (the so-called nicotinic receptors), including those associated with the dopamine reward pathways, which produce sensations of pleasure [F. E. Pontieri, G. Tanda, F. Orzi, and G. di Chiara, "Effects of nicotine on the nucleus accumbens and similarity to those of addictive drugs," *Nature* 382 (1996): 255–57]. Depending on the dose, nicotine can be either a sedative or a stimulant, which means that it would work effectively to counteract anxiety or depression.

37. L. A. Conlay, J. A. Conant, F. deBros, and R. Wurtman, "Caffeine alters plasma adenosine levels," *Nature* 389 (1997): 136.

38. While synthehol is in wide use, Starfleet personnel sometimes opt for the real thing. Aboard the *Enterprise*, Saurian brandy and Romulan ale are brought out on special occasions. And in his bar on Deep Space Nine, Quark tries to develop a market for Tulaberry wine imported from the Gamma Quadrant. For those willing to pay, Quark will also dig into his reserves of Gamzian wine and Maraltian Seer Ale. The Klingons love blood wine and consume it in quantities that rival the most debauched holiday party on Earth. In much greater moderation, the Bajorans enjoy spring wine, which has been compared to French burgundy [U. Tosi, "Trek Wine: To Go Where No Wine Has Gone Before," *Smart Wine Magazine* 1 (January 1997): 56].

39. D. J. Rohsenow and J. A. Bachorowski, "Effects of alcohol and expectancies on verbal aggression in men and women," *Journal of*

Abnormal Psychology 93 (1984): 418–32; J. G. Hull and C. F. Bond, "Social and behavioral consequences of alcohol consumption and expectancy: A meta-analysis," Psychological Bulletin 99 (1986): 347–60.

> > > Chapter 8

1. J. A. Paulos, Innumeracy (New York: Hill and Wang, 1988).

2. The power of wrong-headed ideas is dissected by M. Shermer in Why People Believe Weird Things: Pseudoscience, Superstition, and Other Confusions of Our Time (New York: W. H. Freeman, 1997).

3. The physicist Lawrence Krauss considers these advances from a technological perspective and inquires into their plausibility in The Physics of Star Trek (New York: Basic Books, 1995).

4. As an adult, Julian Bashir is haunted by the knowledge of his genetic legacy. He feels like a fraud. His dark secret finally comes to light when his parents confess what they've done to their son. Bashir is allowed to keep his position as chief medical officer, but his father receives a two-year prison sentence for violating Federation regulations outlawing genetic engineering.

5. In some unfortunate individuals, a recessive genetic mutation produces an unpleasant body odor. As a result of the mutation, the liver cannot break down a protein (trimethylamine, or TMA) that is an ordinary product of food digestion. The excess TMA, which to human noses stinks like rotting fish, finds its way out through the skin, in the person's sweat, and on the person's breath. Typically, a sufferer is isolated socially, and may become clinically depressed or even attempt suicide. Children with this condition are taunted and made fun of. Sometimes, the person reaches adulthood with no idea why he is shunned by others. Recently, the nature of the genetic defect has been unravelled, possibly opening the door to effective gene therapy. (C. T. Dolphin, A. Janmohamed, R. L. Smith, E. A. Shephard, and I. R. Phillips, "Missense mutation in flavin-containing mono-oxygenase 3 gene, FMO3, underlies fish-odour syndrome," Nature Genetics 17 (1997): 491–94.)

6. Actually, thousands of cloned humans are already in our midst. But they don't strike terror into our hearts: they are identical twins, who are essentially the same thing as clones. Michael Quinion provides a short, valuable discussion of the history of the word clone on his World Wide Words Web site at **http://www.quinion.demon.co.uk/words/n-clo1.htm**.

7. In his remarks to the Press Association of Britain, Dr. Ian Wilmut—the human stepfather of Dolly the cloned lamb—addressed the ethical issues raised by genetic engineering: "We believe that it is important that society decides how we want to use this technology [genetic cloning] and makes sure it prohibits what it wants to prohibit.

It would be desperately sad if people started using this sort of technology with people."

8. R. C. Lewontin, *Biology as Ideology: The Doctrine of DNA* (New York: HarperCollins, 1992), 63. Elsewhere, Lewontin points out that with this cloning technique, some tiny fraction of the DNA, 0.06 percent, is not transferred from donor to donee. As a result, Dolly is *not* genetically identical to her donor [R. C. Lewontin, "The confusion over cloning," *New York Review of Books* 44 (23 October 1997), 18–22.

9. Textbook accounts of genetics focus on the genes as the building blocks of life, the blueprints for building an organism. But DNA doesn't just build bodies—it strives to keep them fit. The genes accomplish this by regulating one another's actions based on environmental conditions. Too little circulating cholesterol in the body? This condition triggers DNA-guided production of cholesterol. Too little exposure to sunlight to stimulate adequate levels of vitamin D? DNA-guided vitamin synthesis kicks in and resets the balance. Sensory events as simple as flicking a rat's whiskers can trip a genetic switch that alters the neural connections in the rat's brain. Sequences of DNA are constantly being turned on and off by environmental events we encounter throughout our lives.

10. J. Imperato-McGinley, R. Peterson, T. Gautier, and E. Sturla, "Androgens and the evolution of male-gender identity among male pseudohermaphrodites with 5-alpha-reductase deficiency," *New England Journal of Medicine* 300 (1979): 1233–37.

11. Lewontin, *Biology as Ideology*, 109.

12. U. Bronfenbrenner and S. J. Ceci, "Nature-nurture reconceptualized in developmental perspective: A bioecological model," *Psychological Review* 101 (1994): 568–86.

13. The first appearance of a holodeck in any *Star Trek* television series episode was in an animated episode involving TOS characters (TOS-A, "The Practical Joker"). No holographic facility appeared in the original television series or in the early full-length movies. In *Star Trek: The Next Generation*, our first encounter with a holodeck occurs when the *Enterprise*-D's new first officer, William Riker, enters a simulated parkland full of trees, rocks, and streams. This is where Riker introduces himself to his new colleague, Lieutenant Commander Data (TNG-1, "Encounter at Farpoint").

Thanks to software advances, the *Enterprise*-D's holodeck soon becomes more complex. Its simulations include virtual humanoids who can interact with real crew members in real time. We get our first glimpse of this enhanced capability when Lieutenant Tasha Yar practices her martial arts skills against a simulated opponent (TNG-1, "Code of Honor"). One of *Star Trek's* most memorable holodeck scenes occurs when Lieutenant Commander Data plays five-card stud with three holophysicists, Albert Einstein, Isaac Newton, and Stephen Hawking (TNG-6, "Descent," I). The *Enterprise*-D's crew members use the holodeck for all sorts of things: entertainment (TNG-5, "Cost of Living"), physical exercise

(TNG-2, "Where Silence Has Lease"), reminiscences (TNG-6, "Relics"), strategic planning (TNG-3, "Booby Trap"), recreation of crime scenes (TNG-3, "A Matter of Perspective"), simulation of medical procedures (TNG-5, "Ethics") and job training (TNG-7, "Thy Own Self"). The list of episodes using the holodeck goes on and on. A partial listing can be found in M. Okuda, D. Okuda, and D. Mirek, *The Star Trek Encyclopedia* (New York: Pocket Books, 1994).

14. Besides creating a full-time virtual physician, *Voyager's* holographics systems achieve another first: The ship's bridge houses a holographic projector that makes it possible for individuals at remote locations to appear holographically on the bridge (VOY-2, "Projections"). There, a crew member can carry on conversations with a life-sized three-dimensional hologram. It's no longer necessary to speak to larger-than-life, depersonalized flat pictures on a view screen.

15. B. O. Rothbaum, L. F. Hodges, R. Kooper, D. Opdyke, J. Williford, and M. M. North, "Effectiveness of computer-generated (virtual reality) graded exposure in the treatment of acrophobia," *American Journal of Psychiatry* 152 (1995): 626–28.

16. A. S. Carlin, H. G. Hoffman, and S. Weghorst, "Virtual reality and tactile augmentation in the treatment of spider phobia: A case report," *Behaviour Research and Therapy* 35 (1997): 153–58.

17. K. Glantz, N. I. Durlach, R. C. Barnett, and W. A. Aviles, "Virtual reality (VR) for psychotherapy: From the physical to the social environment," *Psychotherapy* 33 (1996): 464–73.

18. By "telepathy," we mean the transmission of facts, ideas, or emotions from one person to another strictly by the medium of thought. Not everyone agrees with our conclusion that twentieth-century humans are incapable of telepathic communication. The arguments against telepathy are summarized nicely in Martin Gardner's *The New Age* (Buffalo: Prometheus Books, 1988). And those who believe that modern quantum physics supports arguments in favor of telepathy might take time to read J. Casti's *Paradigms Lost* (New York: Avon Books, 1989).

19. Although there are some dissenters from this view, most researchers now believe that the ability to form "theories of mind" is not present at birth, but must be acquired through interactions with the world and with the guidance of parents or caregivers. Imagine a young child watching a puppet show. One puppet—Jerry—puts a toy inside a red box located right in front of the child. Jerry then leaves the stage, and another puppet—George—enters, takes the toy from the red box, and moves it to a yellow box sitting alongside the red one. George now leaves. Jerry comes back and looks at the two boxes, first one, then the other. At this point we ask the child which box Jerry will look in for his toy.

What will the child say? Children under four years old typically answer "the yellow one," even though they have no trouble remembering that George moved the toy. These young children are unaware that their knowledge is not necessarily the same as another individual's knowledge. Somewhere around four years of age, however, this egocentric view changes—they understand that others may have different thoughts from their own.

20. A. S. Lillard, "Other folks' theories of mind and behavior," *Psychological Science* 8 (1997): 268–74.

21. The linguist Deborah Tannen documented the difficulties men and women have in verbal communication [*You Just Don't Understand* (New York: William Morrow and Company, 1990); *Talking from 9 to 5* (New York: William Morrow and Company, 1994)]. According to Tannen, any utterance can mean different things to men and to women, which means we're speaking different languages when we talk to one another. We could also construe this as a difference in the ways in which human males and human females construct "theories of mind."

22. In contrast to Tuvok, *Star Trek's* other famous Vulcan, Mr. Spock, can experience empathy. In *Star Trek: The Motion Picture*, we see Spock shed tears of empathy when he realizes that the alien machine V'Ger is lonely and trying to establish contact with its creator. Of course, Spock is half-human, and it may be his human side that promotes empathy.

23. U. Frith, *Autism: Explaining the Enigma* (Oxford: Blackwell, 1989).

24. J. Chapin, R. S. Markowitz, K. Morxon, and M. A. L. Nicolelis, "Neural population activity in sensory-motor cortex can control an external arm movement system," *Abstracts of Meetings of the Neuroscience Society* 23 (1997): 1400.

Index

Note: the abbreviations "n." and "nn." refer to the Notes section. For example, "218n.13" means "page 218, Note 13," and "220nn.26,28,29" means "page 220, Notes 26, 28, and 29."